*Other Avon books by*
**Isaac Asimov**

# QUASAR, QUASAR BURNING BRIGHT

## ISAAC ASIMOV

 A DISCUS BOOK/PUBLISHED BY AVON BOOKS

Dedicated to
   The Memory of Edmond Hamilton (1904–1977)

The following essays in this volume are reprinted from *The Magazine of Fantasy and Science Fiction*, having appeared in the indicated issues:

AVON BOOKS
A division of
The Hearst Corporation
959 Eighth Avenue
New York, New York 10019

Copyright © 1976, 1977 by Mercury Press, Inc.
Published by arrangement with Doubleday & Co.
Library of Congress Catalog Card Number: 77-82613
ISBN: 0-380-44610-3

First Discus Printing, June, 1979

AVON TRADEMARK REG. U.S. PAT. OFF. AND IN
OTHER COUNTRIES, MARCA REGISTRADA, HECHO EN
U.S.A.

Printed in the U.S.A.

# CONTENTS

# INTRODUCTION

This is my thirteenth collection of science essays taken from *The Magazine of Fantasy and Science Fiction*, and to each one of the previous twelve I have written an introduction.

Twelve different introductions—and now I have to write a thirteenth. The trouble is that it's beginning to seem a dreadful chore to think of something I can say on the thirteenth occasion that I have not said in one or another of the previous twelve. In fact, it even seems an unacceptable task to look over the first twelve books in order to see *what* I had said before.

So I sat down to a cup of coffee which my wife, Janet, had tolerantly prepared for me (she drinks tea) and said, broodingly, "I haven't the faintest idea how to introduce my new collection of science essays."

She said, "Why not write something on what one means by an 'essay'?"

"Great," I said, finishing the coffee fast, and here I am at the typewriter.

For many years now it has been customary for science fiction magazines to include among their stories some non-fiction piece dealing with science. Partly this was for a change in pace and partly, I think, to emphasize the fact that science fiction, at its best, deals seriously with science and that science fiction readers would willingly accept doses of the straight stuff now and then.

Generally, the nonfiction pieces were distinguished from the stories by being termed "articles."

When I first began to write my monthly science piece for *The Magazine of Fantasy and Science Fiction*, ever so many years ago when I was a little over thirty,* I thought

* Oddly enough, I am still a little over thirty today.

of them as "science articles." Gradually, however, there came a shift in my thinking and I began to consider them not articles but "science essays."

The word "essay," as a verb, means "to attempt" or "to try," although it sounds rather antiquated these days and it is rarely used. You "essay a task," for instance.

"Essay" can also be used as a noun, so that you can "make an essay to do something," just as you can "make an attempt," but in this usage the word seems even more antiquated.

In these two senses, the word "essay" is accented on the second syllable.

One of the reasons why "essay" has become antiquated and has fallen out of use is the competing use of the noun "essay" (accented on the first syllable) for the sort of thing you find in this book. It was invented in this sense about 1580 by the French writer Michel Eyquem de Montaigne (1533–92).

He called his short pieces "essays" (*essais*, in French) precisely because he modestly considered them merely attempts to deal with a subject. His pieces were brief and simple, rather than long, detailed, and abstruse discussions. They dealt with some one narrow topic, rather than with a whole field. They were tentative attempts to consider some aspect of a subject after a few hours of musing, rather than being the definitive product of a lifetime of thought.

Most of all, though, *most* of all, an essay is distinguished from more formal expository works by the personal touch. The author does not hesitate to put himself into an essay; in fact, it would scarcely be an essay if he did not. This goes without saying if the essay is subjective and deals primarily with the writer's own thoughts and emotions, but even when the essay is objective and deals with, let us say, a scientific phenomenon, the "I" intrudes and *should* intrude.

It may sound easy to write an essay. You just sit down and maunder about a bit. It doesn't have to be formal because it's *supposed* to be informal. It doesn't have to be very abstruse because it's supposed to be just an easygoing attempt at the subject. And it doesn't have to induce mental cramps through overlong concentration since you are *supposed* to break the spell with asides or little jokes or anything else that occurs to you. What could be easier?

However, it could turn out to be quite hard to do easy things. It takes a long time (and considerable talent) for an actor to learn to act so well that it looks as though he isn't acting. And it takes considerable writing skill to sound as though you're maundering and yet manage to lead the reader to the point.

So if you want to write an essay yourself, you will have to:

1. Have something to say.
2. Cultivate the knack of saying it informally, but *saying* it.
3. Learn to be unself-conscious so that you can get yourself into the essay without blushing or shuffling about uneasily.

Though it's rather troublesome to get the knack of the essay, once you have it, it is just about the most pleasurable writing there is. A novel is a long journey, carefully mapped out freehand; a treatise is a long journey, carefully mapped out with surveyor's tools; but an essay is a pleasant saunter down the road, looking at what happens to lie on either side.

And this book is a series of such things, as my other essay collections are.

I essay the defense of New York City in my own peculiar way—which is to study some interesting statistics about cities.

I essay a celebration of the Bicentennial of the United States by describing the manner in which the nation rose from a rural backwater to the technological leader of the world, and in the process I essay the illustration of one of my favorite beliefs—that *all* significant historical change is nothing more nor less than technological change.

I essay a consideration of why there are Ice Ages and how black holes came to be discovered and what the brightest object in the world is. In the final chapter I even essay an attack (once again and from a new direction) on the foibles of humanity.

But whatever I essay, in one respect the essay succeeds, for in my writing I succeed in enjoying myself, and I hope you enjoy the same success in your reading.

ISAAC ASIMOV
*New York City*

# I

---

# OUR ATOMS

# One
# SURPRISE! SURPRISE!

I've said this before in various places and at various times, but I feel harassed and will say it again. Who knows? Maybe people will eventually believe me.

I am an easygoing person who likes to sit at the typewriter and hit the keys. I work between 8 A.M. and 10 P.M., seven days a week, with frequent interruptions that I try to tolerate. I take no vacations willingly and, except for various biological functions and occasional socializing, there's nothing much besides writing that I'm anxious to do.

Combine that industry (if that's what you want to call it—I've heard it called madness) with an ability to write rapidly and clearly, and the result is an average output of 2,500 words per day (written and published) over a considerable number of years. It's not record-breaking, but it's not bad, either.

But there's no "secret" to this. The industry comes to me without trouble, and I *don't* have to indulge in backbreaking self-discipline. I *like* to write. And as for the ability, why, as far as I know, that I was born with.

Too many people, however, won't accept this and insist there's a "secret" somewhere.

At a luncheon I attended recently, a young man buttonholed me and told me eagerly that he had come to the luncheon precisely in order to meet me. He was a writer who was laboring to change his state of awareness in order to accomplish more and be more like me. Therefore, he said, would I describe to him, in great detail, just how I managed to adjust my own state of awareness.

I said I didn't know exactly what he meant by a state of awareness and I wasn't sure I had one.

He said, "Do you mean to say you are not involved in mind-expansion and altered states of consciousness?"

I shook my head and said, "No."

Whereupon he said, "I'm *surprised!*" and walked away in anger.

But why was he surprised? He was fiddling with himself to become more like me, but I am *already* like me so why should *I* fiddle?

But then, people are often surprised over matters that strike me as not being worth any surprise at all. Let me give you another example, not from my personal life this time, but from chemistry.

We can begin with the periodic table of the elements. This was first worked out by the Russian chemist Dmitri Ivanovich Mendeleev in 1869. Its structure was rationalized by the English physicist Henry Gwyn-Jeffreys Moseley, who devised a way of identifying each element unequivocally by integers ranging from 1 upward (the atomic number).*

In Table 1, I have prepared a form of the periodic table that uses atomic numbers only. Each of the 118 atomic numbers included in the table represents an element, but for the moment, we needn't worry about which name goes with which number. The atomic numbers in the table are divided into seven vertical columns, or periods, which I've numbered, using Roman numerals to avoid confusion with the Arabic atomic numbers.

The number of elements in each period tends to increase as we go up the list. In Period I there are only two elements; in Periods II and III, eight elements each; in Periods IV and V, eighteen elements each; in Periods VI and VII, thirty-two elements each.

It works out this way because of the electron arrangements within the atoms, but that is not something we have to go into in this essay (another subject for another time, perhaps).

The rules worked out from electron arrangement make it possible to go beyond Period VII, in a strictly theoretical way. Thus Periods VIII and IX would contain fifty elements each; Periods X and XI, seventy-two elements each; Periods XII and XIII, ninety-eight elements each; and so on.

* The story is told in some detail in "Bridging the Gaps" and "The Nobel Prize That Wasn't," in my book *The Stars in Their Courses* (Doubleday, 1971).

TABLE 1—THE PERIODIC TABLE

| I | II | III | IV | V | VI | VII |
|---|---|---|---|---|---|---|
| 1 | 3 | 11 | 19 | 37 | 55 | 87 |
| | 4 | 12 | 20 | 38 | 56 | 88 |
| | | | 21 | 39 | 57 | 89 |
| | | | | | 58 | 90 |
| | | | | | 59 | 91 |
| | | | | | 60 | 92 |
| | | | | | 61 | 93 |
| | | | | | 62 | 94 |
| | | | | | 63 | 95 |
| | | | | | 64 | 96 |
| | | | | | 65 | 97 |
| | | | | | 66 | 98 |
| | | | | | 67 | 99 |
| | | | | | 68 | 100 |
| | | | | | 69 | 101 |
| | | | | | 70 | 102 |
| | | | | | 71 | 103 |
| | | | 22 | 40 | 72 | 104 |
| | | | 23 | 41 | 73 | 105 |
| | | | 24 | 42 | 74 | 106 |
| | | | 25 | 43 | 75 | 107 |
| | | | 26 | 44 | 76 | 108 |
| | | | 27 | 45 | 77 | 109 |
| | | | 28 | 46 | 78 | 110 |
| | | | 29 | 47 | 79 | 111 |
| | | | 30 | 48 | 80 | 112 |
| | 5 | 13 | 31 | 49 | 81 | 113 |
| | 6 | 14 | 32 | 50 | 82 | 114 |
| | 7 | 15 | 33 | 51 | 83 | 115 |
| | 8 | 16 | 34 | 52 | 84 | 116 |
| | 9 | 17 | 35 | 53 | 85 | 117 |
| 2 | 10 | 18 | 36 | 54 | 86 | 118 |

Just because we can write numbers indefinitely, following the rules, doesn't mean that it is necessarily useful to do so. In Mendeleev's time, and in ours, too, all the known elements were to be found in the first seven periods. There is, therefore, no practical reason to go higher at the moment.

An important value of the periodic table is that it arranges the elements into groups with similar chemical properties. For instance, atomic numbers 2, 10, 18, 36, 54, and 86 make up the six known noble gases.† Again, atomic numbers 3, 11, 19, 37, 55, and 87 (atomic number 1 is a

† See "Welcome, Stranger," in *Of Time and Space and Other Things* (Doubleday, 1965).

special case) make up the alkali metals,‡ and so on. When a new element is discovered and its atomic number is worked out, it is therefore expected to fit into the table in such a way that its properties are not utterly anomalous. If such an anomaly showed up, the periodic table would be in trouble, but nothing like that has happened as yet.

Up until 1940 only the first six elements of Period VII were known and there was some question as to where they should be placed. To explain the difficulty, let's take a closer look at Periods VI and VII in Table 2. This time I am giving the names of the elements, as well as the atomic numbers. What's more, I am including all the elements now known up to 92, even though two or three were not discovered in 1940 or had just been discovered and were not yet confirmed.

Elements 87, 88, and 89, the first three elements of Period VII, were no problem. They were the certain analogs of elements 55, 56, and 57 in Period VI and belonged right next to them in the table. The problem lay in the three known elements beyond 89. These were thorium (90), protactinium (91), and uranium (92). Where should they be placed?

The point of uncertainty stemmed from the fact that elements 57 to 71, inclusive, in Period VI, make up a group of very similar metals which were commonly referred to as the "rare earth elements."* Chemists had a feeling that the rare earth elements were unique and perhaps a peculiar occurrence within Period VI only. Therefore there was a tendency to skip the positions of the rare earth elements in Period VII and to place thorium (90) next to the first element after the rare earth elements of Period VI, which was hafnium (72). Protactinium (91) would then be next to tantalum (73), and uranium (92) would be next to tungsten (74), and this is shown in Table 2.

Actually, this was wrong. The rare earth elements were not peculiar to Period VI. An analogous group (growing larger and more complicated) must exist in every period thereafter, certainly in Period VII.

‡ See "The Third Liquid," in *The Planet That Wasn't* (Doubleday, 1976).
* See "The Multiplying Elements," in *The Stars in Their Courses* (Doubleday, 1971).

TABLE 2—THE TWO LAST PERIODS

| Period VI | Period VII |
|---|---|
| 55. Cesium | 87. Francium |
| 56. Barium | 88. Radium |
| 57. Lanthanum | 89. Actinium |
| 58. Cerium | |
| 59. Praseodymium | |
| 60. Neodymium | |
| 61. Promethium | |
| 62. Samarium | |
| 63. Europium | |
| 64. Gadolinium | |
| 65. Terbium | |
| 66. Dysprosium | |
| 67. Holmium | |
| 68. Erbium | |
| 69. Thulium | |
| 70. Ytterbium | |
| 71. Lutetium | |
| 72. Hafnium | 90. Thorium |
| 73. Tantalum | 91. Protactinium |
| 74. Tungsten | 92. Uranium |
| 75. Rhenium | |
| 76. Osmium | |
| 77. Iridium | |
| 78. Platinum | |
| 79. Gold | |
| 80. Mercury | |
| 81. Thallium | |
| 82. Lead | |
| 83. Bismuth | |
| 84. Polonium | |
| 85. Astatine | |
| 86. Radon | |

Chemists might have seen that thorium was not particularly similar in its chemical properties to hafnium, or protactinium to tantalum, or uranium to tungsten, but prior to 1940, the chemical properties of these high-atomic-number elements were not really known. It was only beginning in 1940, with the newly discovered uranium fission setting fire to the subject, that the appropriate investigations began to be made.

Then, too, beginning in 1940, elements with atomic numbers higher than 92 were formed in the laboratory, and these were seen to resemble uranium in chemical properties, just as the rare earth elements resembled each other. This meant (as was first pointed out by the American chemist Glenn Theodore Seaborg) that a second set of rare earth elements was present in Period VII after all. The order of the elements was seen to be, therefore, as given in Table 1 and *not* as given in Table 2.

Period VII can now be presented as in Table 3, with the names given of all those elements so far discovered. (Rutherfordium and hahnium are not, to my knowledge, internationally accepted names as yet. The Russians argue the priority of discovery. They call element 104 kurchatovium, for instance.)

The two sets of rare earth elements are now differentiated according to the name of the first element in each. The rare earth elements of Period VI, from lanthanum (57) to lutetium (71), inclusive, are the "lanthanides." The rare earth elements of Period VII from actinium (89) to lawrencium (103), inclusive, are the "actinides."

TABLE 3—THE LAST PERIOD

87. Francium
88. Radium
89. Actinium
90. Thorium
91. Protactinium
92. Uranium
93. Neptunium
94. Plutonium
95. Americium
96. Curium
97. Berkelium
98. Californium
99. Einsteinium
100. Fermium
101. Mendelevium
102. Nobelium
103. Lawrencium
104. Rutherfordium
105. Hahnium
106.
107.
108.
109.
110.
111.
112.
113.
114.
115.
116.
117.
118.

Nuclear physicists have formed thirteen elements beyond uranium (92) and are trying to go still higher in order to see if they can confirm or refute certain theories of nuclear structure they have developed.

All known elements with atomic numbers higher than 83 are radioactive and possess no nonradioactive isotopes. In general, the higher the atomic number, the more intensely radioactive the elements are, the shorter their half-lives, and the greater their instability. The rule, however, is not a simple one. Some elements of high atomic number are more stable than others of lower atomic number.

Thus, thorium (90) and uranium (92) are much more nearly stable than polonium (84). The most stable thorium isotope has a half-life of 14,000,000,000 years and the most stable uranium isotope has a half-life of 4,500,000,000 years so that these elements still exist in the earth's crust in considerable quantities, though they have been slowly decaying ever since the planet was formed.† The most stable polonium isotope, on the other hand, has a half-life of only 100 years. Even californium (98) can beat that, since one of its known isotopes has a half-life of about 700 years.

Nuclear physicists find that they can predict these uneven levels of stability by certain rules they have established concerning proton-neutron arrangements within the atomic nucleus. These rules set up a kind of nuclear periodic table more complicated than the ordinary one of the elements. If these theories are correct, there should be a region of stability in the lower reaches of Period VII, where elements will be found with isotopes possessing unusually long half-lives for such high atomic numbers (see Chapter 2). The presence or absence of such a region will have therefore an important bearing on the theories.

Within this region of stability are elements 112 and 114, so let's see what we can tell about them, if anything, just by looking at the periodic table and using some elementary arithmetic. (Why those two elements in particular? I'll explain later; I promise.)

If we consider 112 first, we see, from Table 1, that it falls just to the right of mercury (80) in Period VI. In fact, it is the as-yet-undiscovered fourth member of the

† See "The Uneternal Atoms," in *Of Matters Great and Small* (Doubleday, 1975).

group whose first three known members are, in order, zinc (30), cadmium (48), and mercury (80). We can call 112 "eka-mercury" by a convention begun by Mendeleev. "Eka" is the Sanskrit word meaning "one" and 112 is the element analogous to mercury in the first period beyond that of mercury.

This group of elements, the "zinc group," shares similar properties. What's more, as in all such groups within the periodic table, certain properties tend to change in some particular direction as we move up the line. Suppose we consider the melting points and boiling points of the zinc group, for instance. This is done in Table 4, where the melting points and boiling points are given in degrees absolute (A), this is, in the number of Celsius degrees above absolute zero, which is −273.1° C. (A Celsius degree is 1.8 times larger than the more common Fahrenheit degree used in the United States.)

In the three known members of the group, the melting point and boiling point go down as the period goes up. It seems fair to conclude that, if the periodic table has validity, the fourth member of the group should have a melting point and boiling point that are lower still than those of mercury.

Can we deduce actual figures? That would be hard to do, since, as we see, the reduction in temperatures is not regular. The melting point of cadmium is 98.5 degrees less than that of zinc, but mercury's melting point is 359.8 degrees less than that of cadmium. That huge drop between cadmium and mercury can't possibly be repeated between mercury and eka-mercury, for that would bring the latter's melting point into negative numbers, which would mean a temperature below absolute zero, which is impossible.

However, in organic chemistry, changes in properties often alternate in character as one goes up the scale of analogs—big change, small change, big change, small

TABLE 4—THE ZINC GROUP

| Period | Atomic number | Element | Melting point (°A) | Boiling point (°A) |
|--------|---------------|---------|--------------------|--------------------|
| IV | 30 | Zinc | 692.5 | 1,180 |
| V | 48 | Cadmium | 594.0 | 1,038 |
| VI | 80 | Mercury | 234.2 | 629.7 |
| VII | 112 | Eka-mercury | ? | ? |

change, and so on. One way of allowing for that is to suppose that since mercury's melting point is a certain fraction of zinc's melting point (comparing two elements two periods apart), then eka-mercury's melting point should be the same fraction of cadmium's melting point. Since mercury's melting point is 0.338 that of zinc, then, if eka-mercury's melting point is 0.338 that of cadmium, it would be about 200 degrees, which sounds reasonable.

Using the same device for boiling points, the boiling point of eka-mercury would be about 550 degrees.

Next, let's consider 114, which is one place to the right of lead (82) in the periodic table as arranged in Table 1 and which we can therefore call "eka-lead." It is the undiscovered member of the group whose six known members are carbon (6), silicon (14), germanium (32), tin (50), and lead (82). The melting points and boiling points of each of those members of the "carbon group" are given in Table 5.

Look at the melting points. There's a big drop between carbon and silicon, a smaller drop between silicon and germanium, a larger drop between germanium and tin, and then actually so small a "drop" that it is a *rise* between tin and lead. Let's, therefore, taken them alternately and compare melting points that are two periods apart:

Carbon/germanium $= 3800/1210 = 3.1$
Silicon/tin $= 1683/505 = 3.3$
Germanium/lead $= 1210/600 = 2.0$

It seems to me, just looking at those figures, that a good ratio for tin/eka-lead would be 2.5. If we divide the melting point of tin, 505 degrees, by 2.5, we get a figure of

TABLE 5—THE CARBON GROUP

| Period | Atomic number | Element | Melting point (°A) | Boiling point (°A) |
|--------|---------------|---------|--------------------|--------------------|
| II | 6 | Carbon | 3,800 | 5,100 |
| III | 14 | Silicon | 1,683 | 2,628 |
| IV | 32 | Germanium | 1,210 | 3,103 |
| V | 50 | Tin | 505 | 2,543 |
| VI | 82 | Lead | 600 | 2,017 |
| VII | 114 | Eka-lead | ? | ? |

TABLE 6—THE NOBLE GASES

| Period | Atomic number | Element | Melting point (°A) | Boiling point (°A) |
|--------|---------------|---------|--------------------|--------------------|
| I | 2 | Helium | 0 | 4.5 |
| II | 10 | Neon | 24.5 | 27.2 |
| III | 18 | Argon | 83.9 | 87.4 |
| IV | 36 | Krypton | 116.6 | 120.8 |
| V | 54 | Xenon | 161.2 | 166.0 |
| VI | 86 | Radon | 202 | 211.3 |
| VII | 118 | Eka-radon | ? | ? |

about 200 degrees for the melting point of eka-lead. Using the same device for the boiling points, we get a figure of perhaps as high as 2,400 degrees for eka-lead.

Let's do one more. Let's try 118, which is the seventh of the noble gases, of which the six known members are helium (2), neon (10), argon (18), krypton (36), xenon (54), and radon (86). Beyond that, 118 would be "eka-radon." The melting points and boiling points of the noble gases are given in Table 6.

In this case, the melting points and boiling points *rise* as one moves up the periods. The rise from helium to neon is 24.5 degrees, from neon to argon is 59.4 degrees, from argon to krypton is 32.7 degrees, from krypton to xenon is 44.6 degrees, and from xenon to radon is 40.8 degrees. Notice the alternation between small rises and large rises. From radon to eka-radon would be a large rise, perhaps about 50 degrees, so that eka-radon's melting point would be about 250 degrees.

The boiling point is always just a little higher than the melting point in the noble gases, but the spread goes up slightly as one goes up the periods. The boiling point of eka-radon might be about 265 degrees.

Now, then, in Table 7, let's summarize the data we have on eka-mercury, eka-lead, and eka-radon. We can give the melting points‡ and boiling points not only on the absolute scale, but on the more familiar Celsius scale, and

‡ A melting point, as temperature is raised, is equivalent to a freezing point as temperature is lowered. I just thought I'd mention it.

TABLE 7—THE EKA ELEMENTS

| Atomic number | Element | Melting point | | | Boiling point | | |
|---|---|---|---|---|---|---|---|
| | | °A | °C | °F | °A | °C | °F |
| 112 | Eka-mercury | 200 | −73 | −100 | 550 | 277 | 530 |
| 114 | Eka-lead | 200 | −73 | −100 | 2,400 | 2,127 | 3,860 |
| 118 | Eka-radon | 250 | −23 | −10 | 265 | −8 | 18 |

the still more familiar Fahrenheit scale. To convert absolute to Celsius, we need only subtract 273. Conversion to Fahrenheit is more complicated, but I'll do that and you won't be bothered.

It would seem, from Table 7, that at ordinary room temperature (sometimes set at 293° A, which is equivalent to 20° C, or 68° F.) eka-radon would be a gas, as are all the other noble gases. It would, however, be far easier to liquefy and to freeze it than is the case with the other noble gases. A cold winter day in New York City would suffice to liquefy eka-radon and a cold winter day in Maine would suffice to freeze it.

Eka-lead and eka-mercury would be liquids at room temperature and, indeed, at any natural temperature that is likely to occur on the surface of the Earth outside Antarctica. A very cold spell in the coldest part of Antarctica might suffice to freeze them.

The two would be quite different with respect to boiling points, however.

Eka-mercury, boiling at 277° C., would be boiling at a low enough temperature to be considered "volatile." Certainly, mercury, which has a higher boiling point, is considered a volatile liquid by chemists.

Mercury has an appreciable vapor pressure, so that in the presence of liquid mercury, there is measurable mercury vapor in the air. This would be even more so for eka-mercury, which would have a higher vapor pressure at corresponding temperatures. In short, eka-mercury would be just like mercury, only more so—with the considerable exception that eka-mercury would be radioactive whereas mercury, as it occurs in nature, is not. (The periodic table of the elements has nothing to say about radioactivity. That is a nuclear property and it is the nuclear periodic table that deals with it.)

Eka-lead, on the other hand, would have a high boiling

point and would not yield any appreciable quantity of vapor in the air. It would be an involatile liquid.

Another sort of property we can deduce from the periodic table relates to the chemical activity of an element; that is, the ease with which its atoms will combine with atoms of another element. As this ease decreases, we can say elements display less and less activity or more and more inertness.

Usually, as one goes up the scale of periods within a particular family of elements, there is a steady trend in the direction of greater activity or of greater inertness. Thus, in the family of the noble gases, the elements grow less inert and more active as we go up the scale of periods. Of the known noble gases, helium is the most inert and least active. Radon is the least inert and most active, and eka-radon, we can be sure, would be still less inert and more active.

These, however, are comparative terms. Radon may be less inert than the other noble gases, but it is still more inert than any of the elements that aren't noble gases, and so would eka-radon be. Eka-radon would still be fairly called an inert gas.

As for the zinc group and the carbon group, its members grow more inert and less active as one goes up the periods. Zinc is a quite active metal; cadmium is less active; mercury is quite inert. The inertness of mercury is obvious; it doesn't rust when it stands exposed to air but remains shiny and metallic; it is too inert to react with oxygen under ordinary conditions. Even when it does react with other elements, the forces holding mercury atoms to the other atoms are relatively feeble and easy to break. In other words, it is easy to get elemental mercury out of its ores, and that is why mercury was one of the metallic elements known to the ancients.

Naturally, we would expect eka-mercury to be even more inert than mercury; it would certainly be an inert liquid.

Carbon, as an element, is fairly inert for a number of good reasons, but it can be nudged into reaction. It will burn in air and it will form a vast number of compounds with other atoms. Silicon resembles carbon in this respect. Germanium is less active and forms compounds less readily, and tin and lead are still less active. Tin and lead are sufficiently inert to hold on to other atoms so weakly as to be easy to isolate. That is why they are two more of the metallic elements known to the ancients.

Eka-lead would be more inert than lead and it, too, would be an inert liquid.

Now that I've taken you this far in my arguments from the periodic table, I will tell you why I have done it. Quite recently calculations have been reported from the Lawrence Berkeley Laboratory, in California, which show that elements 112, 114, and 118, are either gases or volatile liquids and that they are inert.

Apparently those reporting this are surprised; they refer to it as a "striking conclusion."

But again I ask the question, as in my introductory remarks in this essay: Why are they surprised?

The conclusion is not striking at all. I'm sure that the LBL calculations were far deeper, more sophisticated, and more valid than my own attempts to play about with the periodic table. But the results are similar and I would therefore not call the report a "striking conclusion" but an *expected* conclusion.

The September 27, 1975, issue of that excellent periodical, *Science News*, says of the LBL report, "This seems a bit of a surprise because most of the known transuranic elements have been metallic solids."

*Science News* underestimates the situation. *All* the known transuranic elements are metallic solids. Nevertheless, there is no need to be surprised over the presence of inert liquids or gases in positions 112, 114, and 118. Rather, the surprise would have to exist if it were not so, since that would weaken the validity of the periodic table. What's more, the results now reported could have been reached, by using my reasoning in this essay, at any time since 1940, when the correct arrangement of Period VII was pointed out by Seaborg.

In one point, incidentally, I seem to disagree with the LBL report (though I haven't read the original paper and therefore can't be entirely certain).

The second-hand reports I have read seem to indicate that the report claims that eka-lead (114) is a volatile liquid. Well, I admit that 112 (eka-mercury) and 118 (eka-radon) are volatile, but I deny that eka-lead is. Eka-lead is a liquid, yes, but not a volatile one.

If the element is isolated in sufficient quantity in my lifetime to test the matter (I suspect not, alas), then I will be interested to see who is right, LBL or me. I'm betting on me.

# Two
# THE MAGIC ISLE

I was never very good in my lab courses. Whatever gifts I have, none of them includes deftness in experiment work. Teachers who were in any way involved with me discovered this early and reacted to it in different ways.

At one extreme was Charles Reginald Dawson of the Chemistry Department at Columbia University, who supervised my work toward the doctorate. Once, when I had been more than ordinarily undeft in one of my procedures he said to me in the kindly way that never deserted him, "That's all right, Isaac. We'll get someone to do the experiments for you, if necessary. You just keep on having the ideas."*

At the other extreme was Joseph Edward Mayer, also at Columbia, with whom I took a lab course in physical chemistry in 1940. I received a very low mark from him on my report of an experiment on the boiling-point elevation of solutions.

I was not overly surprised at this since my expectations in lab courses were never exuberantly high, but I thought I might as well see Professor Mayer and attempt negotiation. I brought my paper with me and he went over it patiently. I was quite prepared to be told that I had done the experiment sloppily or that I had collected my data thoughtlessly. That wasn't it, however. Professor Mayer looked up at me and said:

"The trouble with you, Asimov, is that you can't write."

For a horrified moment, I stared at him. God knows not everyone likes the material I turn out and a depressingly large number of people say so to my face, but no one has ever seriously told me that I can't write. Except Professor Mayer.

* I have never stopped feeling grateful to Professor Dawson for this and for numerous other kindnesses.

It was an insult I could not abide and I lost all interest in discussing my paper. I gathered up the report and, before leaving, said to him as stiffly and as haughtily as I could, "I'll thank you, Professor Mayer, not to repeat that slander to my publishers."

I passed the course, naturally, but I don't think I ever spoke to Professor Mayer again. Nor, in all the time that has since passed, have I forgotten that remark.

Professor Mayer has had a distinguished career in physical chemistry but his greatest claim to fame is having married a physicist, Maria Goeppert, in 1930. She held on to her name by hyphenating it with his, so that she came to be known as Maria Goeppert-Mayer, and under that name won a share of the 1963 Nobel Prize in physics.

When the announcement reached the papers, my reaction was the simple self-centered one you would expect of a writer. I said, "How do you like that? Goeppert-Mayer has just received a Nobel Prize and yet her husband once told me I couldn't write."

Oh, well, I didn't *really* think her husband's error of judgment ought to have disqualified her, so let's forgive and forget and talk about the work which earned the prize for her.


We'll begin by considering the nuclei of the atoms of the various elements. Each atomic nucleus of a particular atom is made up of a certain number of protons, plus (in the case of every nucleus but the very simplest) a certain number of neutrons as well.

For each particular element, the number of protons in its atomic nuclei is fixed and cannot vary. For instance, all oxygen nuclei have exactly 8 protons. If 1 proton is lost, for any reason, the nucleus is no longer oxygen, but nitrogen. If 1 proton is gained, for any reason, the nucleus is no longer oxygen, but fluorine. The number of protons characteristic of the nuclei of a particular element is the "atomic number" of that element.

The number of neutrons present in the atomic nuclei of a particular element can, however, vary to a certain degree. An oxygen nucleus can contain 8, 9, or 10 neutrons. In each of these cases, the oxygen nucleus that results is stable. That is, left to itself, it will remain unchanged for an indefinite period of time, presumably forever.

These three varieties of oxygen nuclei are "nuclides" and

we can identify them in accordance with the total number of particles—protons plus neutrons—that they contain. We can say that oxygen-16, oxygen-17, and oxygen-18 are the three stable oxygen nuclides, and 16, 17, and 18 are the respective "mass numbers" of those nuclides.

Other nuclides of oxygen are possible. An oxygen nucleus, along with the 8 protons, might have only 7 neutrons or even only 6; it might have as many as 11 or even 12. These nuclides, oxygen-14, oxygen-15, oxygen-19, and oxygen-20 are all unstable, however. If any one of such nuclides comes to exist, it breaks down spontaneously, even when left to itself, and does so in a matter of seconds.

Of course, not all unstable nuclides of the various elements break down in seconds or even in a few years. Some nuclides are not truly stable, but nevertheless endure for billions of years before most of the nuclei break down (see "The Uneternal Atoms," in *Of Matters Great and Small*, Doubleday, 1975). For the purposes of this article, we will consider such nuclides effectively stable since some of them remain unbroken down from their creation eons ago to this day.

The next question is, How do other elements compare to oxygen in the number of stable nuclides they possess?

The answer is that some elements have more nuclides and some have fewer. Let us, however, do a little classification.

We find that elements of odd atomic number, that is, elements with odd numbers of protons in their nuclei, are not remarkable for the number of stable or nearly stable nuclides they possess. Potassium, with an atomic number of 19, has three. All the others have two or fewer.

The situation is quite different with the even-number-proton elements. While the three with the smallest even numbers have only one or two stable nuclides (beryllium has only one and helium and carbon have only two each), all the others, up to and including atomic number 82, have three or more stable or nearly stable nuclides.

In general, then, we can conclude that for a nucleus to possess an even number of protons is a more stable situation than for it to possess an odd number. There are more even-proton nuclides than there are odd-number-proton nuclides, and the even-proton nuclides occur much more commonly in nature. In fact, most of the even-proton nuclides also have an even number of neutrons and the even-proton-even-neutron nuclides make up the bulk of the Universe, if we exclude hydrogen as a special case.[†]

This much I discussed in my article "This Evens Have It" (see *View from a Height*, Doubleday, 1963), but now let's go further. Which element has the greatest number of stable nuclides?

The answer is tin, with no less than ten stable nuclides. Tin has an atomic number of 50 so that it seems that a nuclide with 50 protons possesses so stable a configuration in that respect that the number of neutrons present can vary through an unusually wide set of values without upsetting the stability of the nucleus.

Is there something unusual about the number 50, then? Suppose we consider those atomic nuclei which have 50 neutrons. How many different nuclides, possessing 50 neutrons, are stable? The answer is six, which is unusually high.[‡]

There are thus sixteen different varieties of stable nuclides which possess either 50 protons or 50 neutrons.

Fifty seems such a mysteriously significant number with reference to the stability of nuclear structure that, in 1949, the German physicist J. Hans Daniel Jensen (who eventually shared the Nobel Prize with Goeppert-Mayer) used the term "magic number" in connection with it. In my opinion, that's bad, since the word "magic" should not be used in connection with science, and Jensen later introduced the term "shell number," which is much better. The latter doesn't have a chance, though. Scientists are human and "magic number" is so much more dramatic that even I like to use it.

Are there any other magic numbers? If 50 holds the record for proton stability, what about neutron stability? Is there any neutron number that is represented in more than six stable nuclides? Yes, there are *seven* stable nu-

[†]The simplest hydrogen nuclide, hydrogen-1, has a nucleus made up of a single proton and nothing more. A single-particle is bound to be more stable than any combination of particles so it is not surprising that about 90 percent of all the atoms in the Universe are hydrogen-1, and that the percentage was higher still in the early days of the Universe. In this article we talk of composite nuclei only and, to be sure, in *some* respects *some* composite nuclei are more stable than hydrogen-1 is.

[‡] If you are curious, the six 50-neutron stable nuclides are krypton-86, rubidium-87 (which is slightly unstable), strontium-88, yttrium-89, zirconium-90, and molybdenum-92.

clides, from xenon-136 (54 protons, 82 neutrons) to samarium-44 (62 protons, 82 neutrons), which have 82 neutrons in their nuclei.

What's more, there are four stable nuclides which have 82 protons in their nuclei (representing the element lead). Four may not seem like much, but 82 protons represents very nearly the edge of possible stability. There is only one stable nuclide with 83 protons and none at all that is completely stable yet possesses more than 83 protons (though there are three that are nearly stable). That there are four stable 82-proton nuclides is therefore rather remarkable, and if that is added to the seven 82-neutron nuclides, we might suspect that 82 is a magic number also.

Among the nuclides with fewer particles (where the chances of variation are more limited, in general) there are a surprising number with either 20 neutrons (five of them) or 20 protons (five more) so we might consider 20 a magic number.

Another way of judging stability is by considering the abundance of particular nuclides in the Universe generally. We are not sure exactly how the various nuclides were formed. Presumably the Universe began as a collection of hydrogen-1 nuclides (mere protons) together with perhaps a smattering of simple composite nuclides such as hydrogen-2, helium-3, and helium-4.

Through various nuclear reactions taking place in the core of stars, more complicated atomic nuclei are formed and these are scattered abroad in stellar explosions. In general, the more complicated the nucleus the less abundant it is on a cosmic scale, but this is not a perfectly smooth relationship.

Whatever the manner in which the nuclides are formed, those that are more stable than others are formed more easily and broken up with greater difficulty. They therefore accumulate in greater amounts.

Among the nuclides which occur in the Universe to a distinctly greater degree than other nuclides of similar complexity are the following: helium-4 (2 protons and 2 neutrons), oxygen-16 (8 protons and 8 neutrons), silicon-28 (14 protons and 14 neutrons), calcium-40 (20 protons and 20 neutrons), and iron-56 (26 protons and 30 neutrons).

The matter of abundance is perhaps not a fine enough

test since the presence of magic numbers may not be the only factor involved. Nuclear physicists go at it in another way, too. They check on the readiness with which a particular nucleus will absorb a neutron. The less ready it is to do so, the more satisfied the nucleus is with its existing arrangement and the more likely it is to possess a magic number. Again, some nuclei, if excited and made energetic, will give off a neutron. They will do this most readily if their number of neutrons is just one above a magic number.

Put all the data together and it would seem as if 14, 26, and 30 are not magic numbers and that silicon-28 and iron-56 owe their abundance to other factors. The magic numbers are 2, 8, 20, 28, 40, 50, and 82 for either protons or neutrons. Beyond that the two particles differ. A high magic number for protons is 114; for neutrons, 126 and 184.

Why are those numbers magic? Independently, Goeppert-Mayer in 1948 and Jensen in 1949 worked out a "shell model" of the nucleus and won them their shares of the Nobel Prize in 1963. The protons and neutrons, according to this model, exist in concentric shells, each one larger than the one within. A nucleus is particularly stable if the protons, or neutrons, or, most particularly, both, exist in completed shells or subshells, and that is why "shell numbers" is the better term.

There are two nuclides with the magic number 2: helium-3 and helium-4. Helium-3 has 2 protons and 1 neutron while helium-4 has 2 protons and 2 neutrons. The double helping of magic number makes helium-4 the most stable composite nuclide there is. Of the helium atoms in the Universe only about one out of a million is helium-3, and when a complex nucleus breaks down into something simpler, it often does so by emitting an intact helium-4 nucleus (an "alpha particle"). Indeed, helium-4 is more stable in many ways than hydrogen-1, and it is the tendency to move from hydrogen-1 to helium-4 that powers the stars and, more than anything else, makes our Universe what it is.

There are four stable nuclides which contain either 8 protons or 8 neutrons and of these it is oxygen-16, with 8 protons *and* 8 neutrons, that contains the double dose. In the Universe there are at least three hundred times as many oxygen-16 nuclides as there are of the other three nuclides put together.

There are no less than ten nuclides with the magic number 20. Again the most common of these in the Universe is the one with the double dose, calcium-40, which contains 20 protons and 20 neutrons.

By now, however, a new factor enters. Among the smaller nuclides, the most abundant and, therefore, the most stable, are those with equal numbers of protons and neutrons. The two types of particles do not, however, pack together in precisely equal manner.

All the particles in a nuclide are held together by way of the "nuclear interaction," but whereas there is nothing to counteract this in neutron-neutron or neutron-proton combinations, the story is different in proton-proton combinations.

Between two protons there is a repulsion mediated by the "electromagnetic interaction." This exists only between electrically charged particles, the proton being electrically charged while the neutron is not.

At small distances, such as those involved in the smaller nuclides, the nuclear force is much stronger than the electromagnetic force and the latter can be neglected. The nuclear force, however, fades off rapidly with distance while the electromagnetic force fades off slowly. As the nucleus gets larger, therefore, the electromagnetic force, from end to end of the nucleus, produces a repulsion effect which tends to break up the nuclide and which is, with more and more difficulty, countered by the rapidly weakening nuclear force.

Consequently, as a nuclide grows larger, the number of neutrons it contains must begin to outstrip the number of protons more and more. A greater number of neutrons adds to the nuclear-force attraction, without adding to the electromagnetic-force repulsion.

Calcium-40 is the largest stable nuclide containing equal numbers of protons and neutrons. If we go beyond that, the neutron excess builds steadily. Tin-120, for instance, contains 50 protons and 70 neutrons—a neutron excess of 20. The most massive stable nuclide is that of bismuth-209, which is made up of 83 protons and 126 neutrons for a neutron excess of 43.

Any nuclide which contains more than 83 protons cannot be made stable, apparently, no matter how many neutrons are added. The electromagnetic force will do its work of repulsion and the nucleus will break up sooner or later and fly apart. Three of the known nuclides with atomic numbers of more than 83 are nearly stable and consequently still exist in the Earth's crust. Of these, the most massive is uranium-238 which contains 92 protons and 146 neutrons, for a neutron excess of 54.

If we go beyond calcium-40, then, we can no longer expect double doses of a particular magic number to confer stability. Double doses there may be, but in those cases the protons are at one magic number and the neutrons are at another and higher magic number.

There are ten stable nuclides with the magic number 28 in either protons or neutrons, nine with 40, sixteen with 50, eleven with 82. There is also a nuclide with the magic number 126, which applies only to neutrons.

Of these stable nuclides, exactly three have double doses of magic numbers. They are calcium-48, with 20 protons and 28 neutrons; zirconium-90, with 40 protons and 50 neutrons; and lead-208, with 82 protons and 126 neutrons.

Calcium-48 is not quite stable, but its half-life is in the neighborhood of several tens of quintillions of years, so we might as well consider it stable.

If calcium-48 didn't have a double dose of magic numbers, it is quite likely that it would be distinctly unstable. It is the most massive of the stable calcium nuclides and has a neutron excess of 8. This is an extraordinarily high neutron excess for such a small nucleus. The next larger stable nuclide with so high a neutron excess is nickel-64, with 28 protons and 36 neutrons.

The power of the magic numbers shows up more remarkably if we consider the neutron/proton ratio. In calcium-48, the neutron/proton ratio is 1.4; that is, there are 1.4 neutrons for every proton. Nickel-64 may have a neutron excess of 8, but its neutron/proton ratio is only 1.29. It is not until we reach selenium-82 (34 protons and 48 neutrons) that we reach a neutron/proton ratio higher than that of calcium-48.

Again, though there are ten nuclides that possess either 40 protons or 50 neutrons, only zirconium-90 possesses both. It should not be surprising, therefore, to find that zirconium-90 is, of these ten nuclides, the most abundantly distributed in nature.

That brings us to the final double-dose nuclide, lead-208. Lead-208 is the second most massive of the stable nuclides. It falls short of bismuth-209 by one unit. However, in lead-208 there are 82 protons and 126 neutrons, for a neutron excess of 44, which is 1 greater than that of bismuth-209. It is, in fact, the largest neutron excess among all the stable nuclides.

The neutron/proton ratio is 1.537 for lead-208. This is higher than the ratio of 1.518 which bismuth-209 possesses. Lead-208 does not hold the record for the highest ratio among the stable nuclides. Mercury-204, with 80 protons and 124 neutrons, has a neutron/proton ratio of 1.550. However, mercury-204 makes up only one fifteenth of all the mercury nuclides there are, whereas lead-208 makes up over half of all the lead nuclides. Again, there are well over ten times as many lead-208 nuclides in the Universe generally than mercury-204 nuclides.

Suppose you plot a curve showing the number of protons against the number of neutrons among the stable nuclides. The protons increase steadily as you move upward, the neutrons increase steadily as you move rightward.

Among the simpler nuclides, only those would be stable in which the number of protons and the number of neutrons are equal or nearly equal. We would get a thickish line, then, starting from the origin and making an angle of 45 degrees to the horizontal. As the nuclides grow more complicated, there are present a larger and larger excess of neutrons, so the line begins to curve downward and become more nearly horizontal. Eventually, the line peters out. Even if the nearly stable nuclides are included, the line does not go past the 92-proton mark.

This thickish line of stability is sometimes called the "peninsula of stability" and is pictured as being surrounded by the "sea of instability," which is represented by all nuclides which have too few neutrons or too many neutrons to hold the protons together—or too many protons to be held together by any number of neutrons.

The peninsula of stability can be marked off in a third dimension. We can imagine each nuclide placed at a certain height above the chart, that height being proportional to the extent of its stability as measured by certain of its properties. Naturally, those nuclides which contain a magic number of protons, neutrons, or, most particularly, both, will represent peaks of altitude. Romantic scientists have therefore named certain regions of the peninsula of stability in such ways as "magic ridge" and "magic mountain."

The peninsula is not really solid. For instance, there are no stable nuclides of technetium (atomic number 43) or promethium (atomic number 61). That means the vertical lines representing 43 protons or 61 protons are empty of stable nuclides. I have never heard of these empty lines being named, but I will make so bold as to invent one and call them "proton straits of instability." There are also no stable nuclides with neutrons numbering 19, 35, 39, 45, 61, 89, 115, and 123, and these, by the same token, would represent "neutron straits of instability."

It is interesting that there is no stable or nearly stable nuclide which contains either 61 protons or 61 neutrons, so that 61 seems to be an "anti-magic number" (again my own term).

If we look at the upper end of the peninsula of stability, we see that it ravels off. Beyond the 83-proton mark there is a wide strait of instability, for there are no stable or nearly stable nuclides with proton numbers from 84 to 89 inclusive. There is then one nearly stable nuclide with 90 protons (thorium-232) and two with 92 protons (uranium-235 and uranium-238). We might refer to this sudden emergence from the sea of instability as the "thorium-uranium island" (again my own term).

But what lies beyond uranium? In the last third of a century, nuclear physicists have painfully built up nuclides more complicated than those of uranium, progressing through higher and higher atomic numbers until, as of now, nuclides with as many as 106 protons (and, of course, a considerably higher number of neutrons) have been produced.

On the whole, all of these transuranium nuclides are unstable, though a few of the smaller ones such as neptunium-237 (93 protons and 144 neutrons) and plutonium-244 (94 protons and 150 neutrons) have half-lives in the millions of years. Stability tends to decrease as the atomic number increases. For the really complicated nuclides, the half-lives are a matter of minutes or less.

But in going beyond uranium, we have not yet reached a new magic number. Beyond atomic number 82 (lead) we do not reach a new magic number for protons till we come to 114. Beyond a neutron number of 126 (which is found in lead-208), the next higher neutron magic number is 184. What happens, then, if we reach a nuclide made up of 114 protons and 184 neutrons?

An element with atomic number 114 would be "eka-lead," since it would be just below lead in the periodic table (see Chapter 1) and we are therefore talking of the nuclide eka-lead-298. With a double dose of magic numbers, should not eka-lead-298 be more stable than the other nuclides lying between it and uranium? Even if it were not completely stable, might it not be nearly stable, enough so as to have small quantities still existing in the crust here and there?

Other nuclides in the neighborhood of eka-lead-298 might also be nearly stable so that out in the sea of instability, well beyond the thorium-uranium island, there might be another small island of stability, or a "magic isle" of "super-heavy nuclides."

And now, for the first time, a bit of evidence in favor of the existence of the magic isle has cropped up.

In certain samples of the transparent mineral mica obtained in Madagascar, there are small black discs called "halos." These were first noted in the 1880s and it is now known that they arise from the inclusion in the mica of small bits of radioactive minerals containing thorium and uranium. The thorium and uranium nuclei explode now and then, giving off alpha particles which penetrate the mica for a given distance and discolor it.

The size of most of the halos can be easily matched with the energy and penetrating power of alpha particles from thorium and uranium nuclei. One out of a thousand of the halos, however, is too big. Each would require alpha particles with twice the penetrating power of those that occur in nature.

A group of physicists, headed by Robert V. Gentry of Oak Ridge, Tennessee, speculated that small quantities of superheavy nuclides were the source of the giant halos. They bombarded the giant halos with low-energy protons under circumstances that should produce x-rays on their collision with nuclei. The wavelength of the x-rays would depend on the atomic number of the included nuclei, and if certain wavelengths were detected, that would amount to the finding of superheavy nuclei.

Such wavelengths were indeed detected in tiny amounts, and this *may* represent the first landing by nuclear-physicist Columbuses on the magic isle.

(*Note:* The above essay was written in August 1976. Since then, further investigations by a number of groups have failed to confirm the Oak Ridge findings. It looks, in fact, as though some natural enough misinterpretations were made. The superheavies have *not* been found. That does not mean, however, that they may not be found at some time in the future, either in nature or in the laboratory, and that they may not prove to be surprisingly stable. We'll just have to wait and see.)

# II

---

# OUR CITIES

# Three
# IT'S A WONDERFUL TOWN!

On my eleventh birthday, my father presented me with a copy of the 1931 World Almanac—at my request. Of all the presents I've ever received, I think I remember that most clearly. I read the print and used the statistics to make bar graphs, circle graphs, and line graphs for my own amusement.

With that one book, plus a sheaf of graph paper, a ruler, a compass, and a two-color pencil (red and blue), my parents had me completely out of their hair (except when I was working in the candy store) for at least half a year. You couldn't ask for more out of a total investment of about one 1931 dollar.

I have never recovered from my fascination with almanacs and have just obtained the 1976 World Almanac (to say nothing of the latest Reader's Digest Almanac and CBS News Almanac).

If you go through an almanac carefully and creatively, you can always come out of it with more information than it thinks it is giving you. For one thing, you can always rearrange the information in a new way and make some aspect of its contents more apparent.

And you may be surprised at what you end with. Here, let me show you—

Any almanac will give you the population of the fifty American states and the District of Columbia, sometimes for each of the decennial censuses. Sometimes they will list the states in alphabetical order, sometimes in order of the present population. I consider the second alternative far the more useful.

If you look at a list of the American states in order of population, as, for instance, in the 1970 Reader's Digest Almanac you will see at once that there are seven states, each of which has a population less than that of the District of Columbia. There are forty-two states that have a population less than that of New York City—only eight states, including New York State, with populations larger than that of New York City.

If we consider the population figures estimated for July 1, 1974 (the latest I have), we find that nearly one out of every ten Americans lives in California. We also find that a little over half of all Americans live in the nine most populous states. California has a larger population than that of the nineteen least populous states combined.

But, since this is the Bicentennial Year, let's do something a little more complicated, that involves the magic year of 1776.

When the United States declared its independence, it consisted of thirteen states and had an estimated population of 2,600,000. (The first census was not conducted till 1790, fourteen years later, and at that time the population was 3,929,000.)

Of course, at the time of independence the thirteen states drew their boundaries with a more lavish hand than they do today. Six of them laid claim to lands west of the Appalachians and all the way to the Mississippi. Virginia, in particular, laid claim to the possession of a total area of 354,000 square miles, 40 per cent of the land area of the nation after it won its independence and just about 10 per cent of its present area.

Such claims were given up in the early years of the Republic and are not important. There are, in fact, only three ways in which the boundaries of the thirteen original states were significantly different in 1776 from what they are now.

1. Maryland included the territory now making up the District of Columbia. The District was ceded to the Federal Government to serve as the site of the capital in 1801.

2. Massachusetts included what is now the state of Maine. Maine did not become a state in its own right till 1820.

3. Virginia included what is now the state of West Virginia. The counties of West Virginia seceded from Virginia at the start of the Civil War, or, perhaps more accurately, refused to join the rest of Virginia in seceding from the United States. West Virginia was recognized as a separate state in 1863.

The question is, then, what is the population of the thirteen original states (including the parts that were theirs in 1776) today? It is easily answered (Table 8), but I have never seen a specific table of this sort anywhere else.

As you can see, Virginia and Massachusetts, which were first and second, respectively, in population in Revolutionary times (no wonder they were the political leaders of the colonies) are now numbers 4 and 5, and it is the three Middle Atlantic states that are now in the lead.

The three Middle Atlantic states and Georgia are the only ones of the original thirteen to increase at more than the average rate for the thirteen. Georgia was a frontier state and only a small portion of its present area was settled in 1776; its better than 85-fold increase is understandable.

New York State's nearly 78-fold increase in population is due to the phenomenal growth of New York City. New York City was *not* the largest city in the nation at the time of the Declaration of Independence. That honor belonged to Philadelphia, which had a population of some 33,000 to New York's 25,000. It was the opening of the Erie Canal in 1825 which represented the turning point. After that, trade funneled into and out of New York City on

TABLE 8—THE THIRTEEN ORIGINAL STATES

| State | Population | | Increase ratio 1974/1776 |
|---|---|---|---|
| | 1974 (est.) | 1776 (est.) | |
| 1. New York | 18,111,000 | 233,000 | 77.7 |
| 2. Pennsylvania | 11,835,000 | 298,000 | 39.7 |
| 3. New Jersey | 7,330,000 | 127,000 | 57.7 |
| 4. Massachusetts (and Maine) | 6,847,000 | 328,000 | 20.9 |
| 5. Virginia (and West Virginia) | 6,700,000 | 515,000 | 13.0 |
| 6. North Carolina | 5,363,000 | 270,000 | 19.9 |
| 7. Georgia | 4,882,000 | 57,000 | 85.6 |
| 8. Maryland (and Dist. of Col.) | 4,817,000 | 220,000 | 21.9 |
| 9. Connecticut | 3,088,000 | 212,000 | 14.6 |
| 10. South Carolina | 2,784,000 | 171,000 | 16.3 |
| 11. Rhode Island | 937,000 | 47,500 | 19.7 |
| 12. New Hampshire | 808,000 | 97,000 | 8.3 |
| 13. Delaware | 573,000 | 40,500 | 14.1 |
| Total | 74,075,000 | 2,616,000 | 28.3 |

the way to and from the interior, and in the two centuries of the nation's existence, New York City has increased its population 315-fold, as compared to a 60-fold increase for Philadelphia.

As for the total population of the thirteen states, it now makes up 35 percent of the population of the United States, though in area the states only make up 10 per cent of the nation.

Incidentally, those six of the thirteen original states which were to be among the "slave states" in later American history—Delaware, Maryland, Virginia, North Carolina, South Carolina, and Georgia—had about 49 per cent of the total population of the thirteen states in 1776, but only 34 per cent in 1974. I don't think there's any question but that the social and economic consequences of slavery inhibited their growth.

Having mentioned New York City, I want to turn to the cities of the United States next.

We can almost always find a list of American cities in order of population in the various almanacs, but finding data on their areas is something else. Most people don't think of city areas, because cities are usually represented on maps as dots or little circles. When maps contain city circles in different sizes, or shapes, the more prominent indicators go to the more populous or the more politically significant. Areas are simply never made much of.

Of course, the area of a city is a very artificial thing. A city line can be changed by vote, and suburbs are sometimes brought in for fiscal reasons or sheer boosterism. But then, the mere fact that city lines are artificial, means that population figures are artifical, too. An area just outside a city boundary may be as much a part of the city economically and socially as the area just inside.

Fortunately, the 1970 Reader's Digest Almanac gives figures for the area of the 130 most populous American cities, and we can make good use of this. The area of New York City is, for instance, 299.7 square miles.*

* I should use square kilometers here, but American almanacs still use square miles, and American people still think square miles. I hate to do all the conversions, but you can do so for yourself if you wish. Just multiply any square mile figure by 2.59 and you will have square kilometers. The area of New York is 299.7 × 2.59 = 776.2 square kilometers.

New York City covers a sizable area, and it is almost a quarter as large as the state of Rhode Island. Considering that New York City is 2.3 times as populous as the next most populous city in the United States, it might seem logical to suppose that it is the largest in area, too, but that is not so. There happen to be seven American cities (count them, seven) that are larger in area than New York City.

One of them, many people may recall, is Los Angeles, but Los Angeles isn't the largest city in the United States, either, in terms of area. There are three larger ones, larger by dint of having arbitrarily shoved their boundaries outward in recent years. How many people would guess, off-hand, that the largest city in the United States in terms of area is Jacksonville, Florida. Well, it spreads out over 2.55 times as much area as New York City does.

Here's another table, then (Table 9), of a kind I've never seen before, anywhere. It is a list of the twenty-seven American cities that have an area of over 100 square miles (the Large Cities), taken from the Reader's Digest Almanac list of the 130 most populous American cities. They are presented in the order of decreasing size.

Jacksonville, within its present boundaries, has 63 per cent of the area of the state of Rhode Island. The twenty-seven Large Cities, taken all together, have an area of just under 7,000 square miles, which is about the size of Connecticut and Delaware put together.

TABLE 9—THE LARGE CITIES

| City | Area (sq. m.) | City | Area (sq. m.) |
|------|------|------|------|
| 1. Jasksonville, Fla. | 766.0 | 15. New Orleans, La. | 197.1 |
| 2. Oklahoma City, Okla. | 635.7 | 16. San Antonio, Tex. | 184.0 |
| 3. Nashville, Tenn. | 507.8 | 17. Tulsa, Okla. | 171.9 |
| 4. Los Angeles, Calif. | 463.7 | 18. Detroit, Mich. | 138.0 |
| 5. Houston, Tex. | 433.9 | 19. San Jose, Calif. | 136.2 |
| 6. Indianapolis, Ind. | 379.4 | 20. Columbus, Ohio | 134.6 |
| 7. San Diego, Calif. | 316.9 | 21. Atlanta, Ga. | 131.5 |
| 8. New York, N.Y. | 299.7 | 22. Philadelphia, Pa. | 128.5 |
| 9. Dallas, Tex. | 265.6 | 23. El Paso, Tex. | 118.3 |
| 10. Phoenix, Ariz. | 247.9 | 24. Mobile, Ala. | 116.6 |
| 11. Chicago, Ill. | 222.6 | 25. Huntsville, Ala. | 109.1 |
| 12. Virginia Beach, Va. | 220.0 | 26. Columbia, S.C. | 106.2 |
| 13. Memphis, Tenn. | 217.4 | 27. Corpus Christi, Tex. | 100.6 |
| 14. Fort Worth, Tex. | 205.0 | | |

Of course, most of the Large Cities are located west of the Mississippi, where land was cheaper than in the more settled East. Six of them are in Texas alone. The only two Large Cities in the northeastern region of the nation are New York and Philadelphia. The most populous city that is not a Large City is Baltimore, Maryland, which has a population of 906,000, but an area of only 78.3 square miles.

Obviously, if a city has drawn its lines in a wide sprawl, it may end up including quantities of wasteland, so that it may still have relatively few people within those lines. On the other hand, a city that is tiny in terms of area may nevertheless be well-packed with people.

What we can do is calculate the population density, the number of people per square mile of city area. For instance, Jacksonville has within its 766 square miles, 528,865 people according to the 1970 census, which amounts to an over-all density of 690 people per square mile.† The city of Paterson, New Jersey, on the other hand, has an area of only 8.4 square miles, but its tight-drawn boundary encloses 144,824 people. Paterson holds 17,240 people per square mile. It is twenty-five times as densely populated as Jacksonville.

I will not try, however, to prepare a density table covering all the towns and cities of the United States. I think it will prove a point if I consider the six American cities that are over a million in population (The Great Cities) and list them, not in order of population, but, as in Table 10, in order of population density.

As you see, New York is the most densely populated of the Great Cities. It is 9.3 times as densely packed with people as Houston is. In fact, I strongly suspect that there is no American city of any size that has a population density even close to that of New York City. Surely, one can deduce this if only from the fact that no other city is as packed with high-rise apartments as New York is, so that in no other city do people live in so many horizontal layers over the same patch of ground.

---

† Divide density figures by 2.59 and you'll have the number of people per square kilometer. Thus, the over-all population density of Jacksonville is 690/2.59 = 266 people per square kilometer.

TABLE 10—THE GREAT CITIES

| City | Population (1970) | Area (sq. m.) | Density (per sq. m.) |
|---|---|---|---|
| 1. New York | 7,895,563 | 299.7 | 26,347 |
| 2. Philadelphia | 1,950,098 | 128.5 | 15,175 |
| 3. Chicago | 3,369,359 | 222.6 | 15,136 |
| 4. Detroit | 1,513,601 | 138.0 | 10,968 |
| 5. Los Angeles | 2,809,596 | 463.7 | 6,059 |
| 6. Houston | 1,232,802 | 433.9 | 2,841 |

If we wish to include foreign cities, however, the teeming East has some ant heaps, too, even without the benefit of high-rise apartment houses. Consider Macao, for instance, which is frequently advanced as a case of amazing population density. Macao is a Chinese city, near Canton, which is a Portuguese possession. It has an area of but 5.99 square miles, but crowded within those few square miles are 249,000 people, according to a 1970 census. That makes the population density of Macao about 41,500 people per square mile, which is 1.5 times that of New York.

We are talking now of over-all population densities, however, and within any city there are always relatively crowded areas and relatively empty ones. It may not always be easy to parcel a city into discrete units that make sense in order to compare densities, but in the case of New York City, there is no problem.

New York City is made up of five boroughs, each of which is a separate county of New York State. The separate boroughs—Manhattan, Bronx, Brooklyn, Queens, and Staten Island—are quite familiar to the rest of the nation since they are so frequently mentioned in books, plays, motion pictures, songs, and so on.

Each of the boroughs is, in itself, a well-populated area. Four of them, in fact, would be among the Great Cities, if each were counted separately, and one would be a Large City. Table 11 shows what a list of the American Great Cities would be like if the five boroughs of New York City were separate cities.

Next, let's confine ourselves to the boroughs and prepare Table 12, showing their population densities.

As you can see, the island of Manhattan has an over-all population density about 1.7 times that of Macao and

TABLE 11—THE GREAT CITIES (NEW YORK BROKEN UP)

| City | Population (1970) |
|---|---|
| 1. Chicago | 3,369,359 |
| 2. Los Angeles | 2,809,359 |
| 3. Brooklyn | 2,602,012 |
| 4. Queens | 1,987,174 |
| 5. Philadelphia | 1,950,098 |
| 6. Manhattan | 1,539,233 |
| 7. Detroit | 1,513,601 |
| 8. Bronx | 1,471,701 |
| 9. Houston | 1,232,802 |

TABLE 12—THE BOROUGHS OF NEW YORK

| Borough | Population (1970) | Area (sq. m.) | Density (per sq. m.) |
|---|---|---|---|
| 1. Manhattan | 1,539,233 | 22 | 69,965 |
| 2. Brooklyn | 2,602,012 | 70 | 37,171 |
| 3. Bronx | 1,471,701 | 40 | 36,793 |
| 4. Queens | 1,987,174 | 105 | 18,925 |
| 5. Staten Island | 295,443 | 63 | 4,690 |

maintains that density over an area 3.67 times that of Macao.

Another way of looking at it is this. Manhattan is the smallest county in the United States. The nation's largest county is San Bernardino, California, which has an area of 20,119 square miles. This is 914 times as large as Manhattan and is almost as large, in fact, as the state of West Virginia. That enormous county, however, is mostly the Mojave Desert and its total population is 681,535, less than half that of the tiny island of Manhattan.

Furthermore, let's not base the population density of Manhattan on its *residents* (of whom my wife and I are two)—the dead of night density. During the day, people flood into Manhattan from the outlying boroughs, from Westchester, Long Island, New Jersey, and Connecticut. I suspect that the population density of Manhattan at noon surpasses the 100,000-per-square-mile mark.

A density of 100,000 people per square mile is a hard thing to visualize. If the state of Delaware (second smallest in the nation) were packed with people at this density, it would contain the entire population of the

United States. If the state of Kentucky were packed with people at this density, it would hold every man, woman, and child on Earth.

I doubt that there is anywhere in the world a place where the population density is higher than in Manhattan at noon under ordinary day-to-day conditions, or *can* be, at the present level of technology. If there are portions of Tokyo or Shanghai (the only two cities with populations greater than New York) with higher population densities, then we might ask if they are accompanied by as high a standard of living as exists in Manhattan.

In this sense, the island of Manhattan is the most amazing production of the human species. Nowhere on the face of the Earth, either now or ever, has so high a population density been supported over so sizable an area at so high a standard of living. And, physically, nowhere on the face of the Earth, now or ever, has anything been constructed by man to equal the skyline of Manhattan—that vast complex of enormous and intricate structures. By comparison, the Pyramids and the Great Wall are just what they are— huge heaps of dead rock.

Consider, too, that, for a variety of historic reasons, New York City has drawn to itself an incredible diversity of people, languages, and cultures—again to a point never seen elsewhere, either now or in the past. The great cities of ancient times were only foreshadowing whispers, even the greatest of them. And as for the populations of Tokyo and Shanghai, which are greater than New York in terms of numbers, they are each a homogeneous mass—one language, one culture. Only New York comes near to holding within itself all the variegated splendor of humanity.

New York City, the musical comedy title says, "is a wonderful town." Yes, it is—but it may be dying.

New York City is not a nation, so it can't police its boundaries. It can't prevent the affluent leaving for the suburbs or for California. It can't prevent the indigent from entering.

Many thousands of immigrants have entered the United States through the port of New York, have been helped, educated, given jobs, introduced to the American way of life. The crowds, the cold-water flats, the sweatshops, weren't nirvana, but where would it have been better? The immigrants made their way and their children and grandchildren did well—and left New York for greener places.

The Golden Door was closed to all but a trickle of immigrants, half a century ago (just one year *after* I myself arrived), but there are now many thousands of "immigrants" reaching New York from other parts of the nation. They, too, are indigent; they, too, need help; but now New York is in financial trouble and it cannot help them, and no one wants to help it help either. It can no longer be the gateway to the American dream—and people laugh at it for trying. "Fiscal irresponsibility," it is called.

What hurts is that I'm afraid that among those who laugh and sneer at New York are the descendants of some of the Europeans who learned how to be Americans in New York—mockers who now feel no need to repay or to pass on to others the good their parents and grandparents have received.

New York is not alone in the nation in the miseries that the changes since World War II have brought to it. It is the largest city, the one with the softest heart, and therefore the one that makes the best target—but make no mistake, it is the head of the spear. Where it goes, the rest of the nation will follow. If the nation is to be saved, New York must be saved.

Let's consider the new wave of immigrants that have entered New York City since World War II. A large percentage of them are Blacks and Hispanics, looking for a better life now, as my parents did half a century ago.

New York is now the largest Black city in the world. Africa's largest Black city is Kinshasa, Zaire, which has a population of 1,623,760. New York City, however, contains 1,666,636 Blacks. (Mind you, I don't trust either figure to the last digit or even to the nearest ten thousand, but that's what the censuses say.)

New York has been steadily Blackening since World War II, and so has every other large city in the United States. Let's consider (in Table 13) those cities which contain more than 100,000 Blacks (twenty-five of them) and list them in order of percentage Black as of 1970 and compare that with the percentage Black of 1960. I've never seen

TABLE 13—BLACK PERCENTAGES

| City | Black population (1970) | Per cent Black 1970 | Per cent Black 1960 |
|------|------------------------|---------------------|---------------------|
| 1. Washington, D.C. | 537,712 | 71.1 | 53.9 |
| 2. Newark, N.J. | 207,458 | 54.2 | 34.0 |
| 3. Atlanta, Ga. | 255,051 | 51.3 | 38.2 |
| 4. Baltimore, Md. | 420,210 | 46.4 | 34.7 |
| 5. New Orleans, La. | 267,308 | 45.0 | 37.2 |
| 6. Detroit, Mich. | 660,428 | 43.7 | 28.9 |
| 7. Birmingham, Ala. | 126,388 | 42.0 | 39.6 |
| 8. Richmond, Va. | 104,766 | 42.0 | 41.8 |
| 9. St. Louis, Mo. | 254,191 | 40.9 | 28.6 |
| 10. Memphis, Tenn. | 242,513 | 38.9 | 37.0 |
| 11. Cleveland, Ohio | 287,841 | 38.3 | 28.6 |
| 12. Oakland, Calif. | 124,710 | 34.5 | 22.8 |
| 13. Philadelphia, Pa. | 653,791 | 33.6 | 26.4 |
| 14. Chicago, Ill. | 1,102,620 | 32.7 | 22.9 |
| 15. Cincinnati, Ohio | 125,070 | 27.6 | 21.6 |
| 16. Houston, Tex. | 316,551 | 25.7 | 22.9 |
| 17. Dallas, Tex. | 210,238 | 24.9 | 19.0 |
| 18. Jacksonville, Fla. | 118,158 | 22.3 | 52.6 |
| 19. Kansas City, Mo. | 112,005 | 22.1 | 17.5 |
| 20. New York, N.Y. | 1,666,636 | 21.2 | 14.0 |
| 21. Pittsburgh, Pa. | 104,904 | 20.2 | 16.7 |
| 22. Indianapolis, Ind. | 134,320 | 18.0 | 20.6 |
| 23. Los Angeles, Calif. | 503,606 | 17.9 | 13.5 |
| 24. Boston, Mass. | 104,707 | 16.3 | 9.0 |
| 25. Milwaukee, Wis. | 105,088 | 14.7 | 8.4 |

a table quite like this anywhere, but I'll prepare it for you out of 1976 CBS News Almanac data. (And note that the percentage of Blacks in the nation as a whole is about 11.1.)

The only two cities in which the Black percentage dropped during the 1960s are Jacksonville and Indianapolis. Both cities, however, enlarged their areas in the last decade, I believe, bringing in largely White suburban areas, so that the figures for the two years are not comparable.

Elsewhere, we can see that the Black percentage is going up rapidly, partly because Blacks are moving into the cities from rural areas and partly because Whites are moving out of the cities into the suburbs. As a matter of fact, the White efflux is often greater than the Black influx, so that the population of some of the large cities of the United States is actually dropping despite the continuing population increase in the nation as a whole.

As examples, between 1950 and 1970, Cleveland's population dropped from 915,000 to 751,000 and Boston's from 801,000 to 641,000. This is a population of 324,000 for the two cities, even while the nation's population went up 52,000,000 in those two decades.

Since, as it happens, it is the prosperous who move out of the cities and the indigent who move in, most American cities find that less money can be raised by taxes and more money for salaries, services, and welfare must be spent.

Where is the money to come from?

Perhaps from nowhere. Rural areas and small towns have traditionally distrusted and disliked the cities, and since small towns and rural areas have always been proportionately overrepresented in the various state legislatures, and even in Congress, cities routinely get the back of the hand from the states and nation.

The rural areas and small towns have been playing less and less of a role in American life, to be sure. In 1776 some 95 per cent of the entire national population was rural, whereas in 1960 only 30 per cent of the population did not live in or near a city of over 50,000 population. In 1970 the rural figure had dropped further to 26.5 per cent.

The cities, however, did not gain as the rural areas lost. It was the suburbs that gained—the affluent suburbs, which depend on the cities economically but, Pilatelike, wash their hands of urban troubles.

Increasingly, though rarely mentioned out loud, the division between suburb and city is the division between White and Black, between Anglo and Hispano, between rich and poor. It is not considered polite any longer in the United States to say nasty things about Blacks and Puerto Ricans and the poor (except among friends, of course), but it serves just the same purpose to say nasty things about the cities, and that can be done out loud in the best circles.

And one city in particular, in the face of nationwide changes (and even world-wide changes, if we're going to talk about energy shortages and inflation), has foolishly attempted to maintain the standards of a more idealistic day. It has tried to be more generous to its poor and its employees than others have and to offer more services than others do. Naturally, it went broke in the process and the President of the nation makes political hay out of encouraging people to laugh at New York. He makes fun of the city at home and abroad, and refuses to help.

What the heck, *he's* all right; the leak isn't at *his* end of the boat.‡

‡ Yes, it was. This essay was written in November 1975. One year later, in November 1976, the President, Gerald Ford, ran for re-election. He did poorly in New York City. Jimmy Carter's margin in New York City gave him New York State, and New York State's electoral votes defeated Ford.

# III

---

# OUR NATION

## Four
# MAKING IT!

I know a woman who is the daughter of a very well-known scientist, but who grew furious when I once introduced her as that daughter. She wanted to make it on her own—and she had every chance to do this since, before marriage, her last name was a common one, and, after marriage, it was utterly changed. It was only necessary for no one to snitch.

My own beautiful, blonde-haired, blue-eyed daughter (who is currently a college sophomore and unmarried) doesn't quite have this chance because her last name is a dead giveaway. Fortunately, she doesn't mind, partly because she is rather fond of me and partly because the relationship is an excellent icebreaker when she meets new people. In fact, she has reduced the whole thing to a fine art. She called me long distance (collect, of course) recently to tell me of a particularly spectacular instance.

The scene is somewhere in Harvard Square. My daughter, Robyn, and a couple of friends are exchanging pleasantries with (I presume) some Harvard students and names are exchanged:

ROBYN (*pleasantly*): And I am Robyn Asimov.
YOUNG MAN (*in growing agitation*): Say, you're not going to tell me you're related to Isaac Asimov. He's not your uncle, is he?
ROBYN (*scornfully*): Of course he's not my uncle.
YOUNG MAN (*slowly deflating*): Oh.
ROBYN (*carefully waiting for the moment of complete relaxation*): He's my father.

Robyn refused to try to describe the explosion that followed, for lack of adequate words, but assured me that it was very satisfactory. Naturally, I laughed, for Robyn has precisely my own distorted sense of humor, and I said, fondly, "I guess you're my daughter, all right." —Not that there was ever even the slightest scintilla of doubt on that subject.

But with that thought in mind, I would now like to go back to Revolutionary days and point out that the United States is technologically Great Britain's daughter and that, despite rebellion on one side and disinheritance on the other, the relationship shows. Let me explain—

Two hundred years ago, when our nation declared its independence, it was not an industrially developed society even by the standards of the time. It was almost entirely rural and anything that required the least bit of sophistication in manufacture had to be brought in from outside.

In fact, the American colonies were deliberately kept underdeveloped by Great Britain, which wanted those colonies to serve as a source of raw materials (bought cheaply by herself only) and as a market for finished products (sold dearly by herself only). In this way, Great Britain could profit at the expense of the colonies, and, of course, the colonials grew increasingly indignant over the matter.*

The colonials carried on an economic struggle against Great Britain as a result, first by smuggling, then by boycotts, and finally, when Great Britain was stung into rigor, by force of arms. (I don't want to seem cynical, but the chief "liberty" at issue between the mother country and the colonies was the liberty to make money and who should have it—the British merchants and landowners or the American merchants and landowners? Part of the fallout of this economic struggle was the series of other liberties embodied in the Bill of Rights, however, and for that I am grateful.)

When, in 1783, the British government was finally, and most reluctantly, compelled to recognize the independence of the United States, they were left with no great feeling of love for us.

British enmity led them to find reason to retain certain bases on American soil; to arm, clandestinely, the Indians of our northwest frontier; to hamper our trade in a hundred ways. They might have continued nibbling away at us in a continuation of the war by economic means until

---

* See Note at the end of this chapter.

55

they brought us down or broke us up into several even weaker and more dependent regions were it not for the French Revolution and the coming of the Napoleonic wars. With the greater danger facing them just across the Channel, the British let us go.

The most subtle and dangerous aspect of British enmity lay in the technological sphere, however, and this is rarely mentioned in the history books. Consider—

Great Britain, even while the American Revolution was being fought, was industrializing itself in a new way, one that the world had never before seen.

British industry was taming inanimate energy and turning it to the work usually done by human muscle in a highly useful way. There had been, to be sure, "prime movers" (machines for the putting of inanimate energy to work), based on wind and water, ever since men had used sails, water wheels, and windmills—but in 1769, the Scottish engineer James Watt designed a steam engine that was an improvement over older models and was the first practical prime mover based on the heat produced by burning fuel.

Fuel could be burned anywhere, so that energy did not depend on place, as did water power, which could be used only at certain sites on certain rivers. Fuel could be burned anytime, so that energy did not depend on the whim of nature, as when wind blew or did not blow. Fuel could be burned in any quantity so that man's needs were not dictated to by the accident of water and wind capacity at any given moment.

By 1774, just on the eve of the American Revolution, Watt went into partnership and began to produce steam engines commercially. In 1781, when the Battle of Yorktown finally decided the struggle in America's favor, Watt devised mechanical attachments that ingeniously converted the back-and-forth movement of a steam-driven piston into the rotary movement of a wheel, and by one type of movement or the other, the steam engine could then be made to power a variety of activities. Almost at once, for instance, iron manufacturers were using it to power bellows to keep the air blasts going in their furnaces and to power hammers to crush the ore.

The vital next step was taken by Richard Arkwright, born in Preston, Lancashire, on December 23, 1732, as the youngest of thirteen children. He was, in his youth, a barber and wigmaker and laid the foundation of his fortune with a secret process for dyeing hair. In 1769 he patented a device that would spin thread by mechanically reproducing the motions ordinarily made by the human hand.

Of course, it wouldn't do very much good to have a machine imitate the hand if a hand had to direct the machine. The machine could, however, be powered by something less skillful than the human hand. Arkwright first used animals to power his mechanical spinner and then water power. In 1790 he began to use the steam engine, and that was the crucial turning point. The modern factory was born, and when Arkwright died in 1792, he was a millionaire.

The factory system, in its early days, had its bad points. Workers were thrown out of work at a time when society felt no responsibility for them—but merely hanged those who stole bread to feed starving children. And since what human supervision of machinery was required involved neither experience nor strength, children were employed because they would work more cheaply. The crimes committed against children in the early years of the factory system will not bear repeating—at least by me. The only thing more incredible than the cruelty with which they were treated was that so many fundamentally decent people managed to look the other way. Yet the flaws were eventually corrected, and the benefits remained.

As other aspects of the textile industry were also mechanized, cloth could be produced in such quantities and so cheaply that a much larger percentage of the human race could be decently clothed than had ever been possible before. Since clothes and increasing quantities of other "consumer goods" produced by the burgeoning factories had to be sold to ordinary people, these began to be viewed as "customers" and, as customers, they were more valuable than they had been as "yeomen" and "varlets," so that Great Britain perforce moved in the direction of democracy.

Great Britain's possession of coal to fuel her engines, ships to transport her goods, and the experience required to build and broaden her industrialization made her the richest and most powerful nation in the world. She held that position throughout the nineteenth century and in its course became the greatest empire builder (at the expense of nonindustrialized people) that the world had ever seen.

The industrialization of Great Britain, coming just as the United States had broken loose, threatened to abolish utterly the gains the colonists had thought they had made in winning their "freedom." What freedom would they have if Great Britain could make clothing in such quantities and of such quality that the American home products could not even begin to compete? The United States would be forced to sell cotton (and other raw materials) to the British at their price and buy the cloth (and other finished products) from the British, again at their price. With Great Britain setting the prices, we would lose and they would gain.

That was what the British had wanted before the Revolution and that was what they could have after the Revolution. That is how the business of colonialism works, whether the colony openly belongs to the nation that exploits it or whether it pretends to be self-governing.

The only way out was for the United States to develop a textile industry of its own. But how? The United States had its ingenious individuals, of course—there had been Benjamin Franklin, for instance†—but sheer ingenuity was not likely to do the job fast enough. British secrets had somehow to be stolen if the proper speed were to be attained.

Mind you, that wasn't easy. Great Britain knew perfectly well that its wealth and strength depended on its maintaining and, if possible, extending its industrial lead over the rest of the world, and it made every effort to do this. Blueprints of the new machinery were not allowed to leave the country, and neither were engineers who were expert in the new technology. And it was only logical that the British were determined that the Americans, above all, were not to get it.

The new textile machinery was to the British of 1790 what the nuclear bomb was to the Americans of 1945, as far as fear of disclosure was concerned. And, on the other hand, the Americans of 1790 were as avid to learn of the new textile machinery as the Soviet Union of 1945 was avid to learn of the nuclear bomb.

† See "The Fateful Lightning," in *The Stars in Their Courses* (Doubleday, 1971).

The United States did what you would expect it to do under such conditions. It did its best to find defectors— just as the Soviet Union did a century and a half later.

This brings us to Samuel Slater, born in Belper, Derbyshire, on June 9, 1768. He served as an apprentice to a partner of Richard Arkwright. He worked with the textile machinery and knew it intimately. However, Great Britain was a class-ridden society, and upward mobility being difficult to manage, Slater knew his advance would be limited.

(Arkwright, to be sure, from an insignificant start, had attained great wealth and a knighthood, too, but that was the exception. In fact, exceptions of this kind do harm, for they help to maintain an unjust system by supplying the gloss that hides the injustices. The success of one is used to justify and obscure the oppression of ten thousand.)

It seemed to Slater that he could do better across the sea where a young and even chaotic society left wealth and prestige up for grabs—all the more so since the United States was offering bounties (that is, bribes) for the kind of knowledge he possessed.

Slater couldn't take any blueprints with him, of course, so he painstakingly went about memorizing every detail of the machinery; there was, after all, no way for the authorities to search the possessions of his mind. Nor could he emigrate as an engineer, so he disguised himself as a farm laborer and sneaked out of the country.

He "defected," actually. What else can you call it?

In 1789 he arrived in New York and then made contact with the Browns, the richest merchant family in Rhode Island. (Brown University is named for Nicholas Brown, whose money led to the original foundation, and Slater dealt with Moses Brown, Nicholas' son.) By 1793 Slater, working from memory, built in Pawtucket the first factory based on the new machinery that the United States was to see within its borders. He went on to build other factories in New England.

This was just a beginning—but so was the Declaration of Independence just a beginning. Slater's beginning burgeoned as the Declaration's beginning did, and the end of it was that the United States became an industrial power.

If George Washington was the Father of his Country, Samuel Slater was the father of the industrialization of his adopted country. However, politics and war are glamorous and economics is usually thought to be dull,‡ so while George Washington is all-pervasive in this country, and never more so than during the Bicentennial year, Samuel Slater is virtually unknown, though his deeds gave the United States a greater chance at true independence than Washington's, economically unsupported, might have done.

There is, to be sure, a Slaterville on the north-central border of Rhode Island, which is named for Slater, but I can't help wondering how many of the inhabitants of the village know who it is named for, and why.

As a result of the Industrial Revolution, which came to the United States in 1793, our country made it. It was not to be a colony of Great Britain in any way, and, having gained its political independence, it could strive for and gain economic independence as well.

Of course, *within* the country there existed the problem for some sections that the nation as a whole had avoided. New England and, to a lesser extent, other northern states were industrialized, while the southern states, enamored of a gracious, chivalrous existence (for a small percentage of the population) on the backs of slaves rather than machinery, remained rural.

It was a fearful mistake for the southern states, for they became, in effect, colonies of the industrialized northern states, and particularly of New England. The New England cotton mills bought the South's raw cotton cheap and sold it cloth dear, and tariff walls were set up to keep the southern states from finding better deals elsewhere. The gracious chivalry of the plantation owners did not keep them from being in hock to the northern capitalists right up to their gracious and chivalrous ears.

Slavery was an emotional issue in the United States of the 1850s as the rights of Englishmen was an emotional issue in the colonies of the 1760s, but it was economics that produced the real hostility in both cases. Because the

‡ Far be it from me to be censorious here. I have written a dozen history books in which I deal with politics and war at great length and with economics only briefly, precisely because the first is glamorous and the second dull, so you understand I don't blame others for doing the same. I just point it out—that's all.

northern states industrialized themselves, they grew prosperous and populous and the United States Government, run by the centers of population (who had the votes, you see) naturally organized itself in such a way as to further favor the already favored.

The southern states found themselves sliding ever back into a colonial position that was clearly to be permanent unless they did something drastic. They tried to increase their power by expansion at the expense of Mexico (over considerable northern objection), and when that didn't work, they decided to leave the Union and form a Confederacy of their own.

The southern states never recognized (or wouldn't admit) that it was their own deliberate choice to be a slave society rather than a machine society that was holding them back, so they never realized that they couldn't win. Even if they had managed to establish their "independence," they couldn't win.

If victory could be determined in a purely military fashion on the battlefield, the new Confederacy had a good chance actually. Although the Confederate population was substantially lower than that of what remained of the Union, that was not really a decisive factor.

The Confederacy had the best generals (Robert E. Lee was, beyond any doubt, the greatest Captain ever born in the territory of the United States), the best cavalry, the best soldiers. And they had the advantage of the defense.

Whereas the Union armies had to advance and occupy the Confederate territory (which was large and had few vital centers) against desperate resistance if they were to win the war, the Confederate armies had only to hold the Union to a stalemate. They did not have to invade the northern states; they were not fighting for territory. They could even afford to retreat and give up some of their own territory. All they had to do was hang on, in any way and however precariously—just hang on till the Union grew sick of the mess and gave up.

We know what that's like now. We would have had to destroy the Vietcong and North Vietnamese in the late unlamented unpleasantness in order to win, but they did not have to destroy us. They had no way in the world of doing more than pinpricking us. They, however, didn't have to win in the conventional way; they just had to hold on at any cost till we were tired of the whole miserable business. They did—and won a war in which they didn't win a single battle in the field.*

Then, too, a century ago, the Confederacy expected to have industrial nations of Europe on their side, especially Great Britain. The reasoning was that Great Britain would need the cotton of the Confederacy for their textile factories, and rather than risk their own economic ruin, the British (so reasoned the Confederacy) would break the federal blockade and take up the role of arsenal for the slavocracy.

Neither calculation worked. To take up the latter first, Great Britain did *not* side openly with the Confederacy. The British ruling classes would have liked to, if only to weaken the United States and throw the Americans open to British exploitation all the more, but they never went past the point where their help to the Confederacy would be decisive. (Part of the reason for this was that the very British textile workers who were thrown out of work when the factories shut down for lack of cotton rallied in great demonstrations against the Confederacy that could have put them back to work because of their disapproval of slavery. It was an example of something that happens but rarely in human history—the victory of long-term idealism over short-term advantage.)

And why didn't the Union weary of the Civil War, as a century later, the United States wearied of the Vietnam War? Of course, the Civil War was a lot closer to the heart—a *lot* closer. Another factor was that the Union had the incredible luck to have for its President Abraham Lincoln—who never gave up.

Lincoln had his goal, and although he was laden with recurrent disaster on the battlefield, with stupidity, corruption, and near-treason at home, and although he bore more on his shoulders than any one man should have had to bear,† he never wavered, he never gave up, he never for one moment lost sight of where he was going and why, nor, in the process, did he lose a bit of the saving humor in his mind nor a drop of the milk of human kind-

---

* Even the Tet offensive had been a tactical defeat for them.

† After the Battle of Fredericksburg on December 13, 1862, the most disastrous of all the Union defeats—brought on entirely by incapable generalship—Lincoln said, "If there is a worse place than hell, I am in it."

ness in his heart—but I've got to get off the subject, or I'll never end this sentence.

Yet suppose Lincoln had broken and suppose the British had come in, and suppose the Confederacy had dictated a peace that left them independent, with Great Britain ready to guarantee that independence against a later American attempt to defeat them. Would the Confederacy have won?

Not at all. They would have gained nothing. As long as they remained an essentially rural economy based on slave labor, they would have remained a colony. All they would have accomplished would have been a change of masters—no more New England, but Old England. And Great Britain would have insisted on freedom for the slaves, too.

—But never mind. The Confederacy didn't win. The Union won the war. Why?

It wasn't because Great Britain did not intervene. That only gave the Union a chance not to lose. It wasn't because Lincoln was Lincoln; that only meant the Union wouldn't give up. What made them *win?*

To see why that was, let's go back once more to the early days of the Republic.

In the day when the United States was winning its independence, Frederick II was on the throne of Prussia. He had won several wars against the larger monarchies that surrounded him and he is therefore commonly known as "Frederick the Great"—the last monarch in history to receive that title. In 1783, when American independence was confirmed and a fact, Frederick was seventy-one years old, had been king for forty-three years, and had only three years left of life. He was Europe's elder statesman and was anything but a fool—and he had a clear opinion of the new nation. He said it would not survive.

His reasons were logical. The new country called the United States was too large and too empty to hold together. It was a reasonable opinion for a monarch to hold whose own nation was small and who was surrounded by many other small nations. Smallness was, to him, a fact of life. And he may even have been correct if his unspoken assumption, that the state of technology would remain unchanged, had been correct.

If we look at the early United States without the benefit

of hindsight, we can see where Frederick might have had reason on his side. The United States was a collection of thirteen different states, each jealous of its sovereignty and each by no means unsuspicious of the rest. They couldn't possibly pull together over the long run.

The Constitution meant a change for the better, for it produced a federal government superimposed above the states through a voluntary surrender by those states of parts of their sovereignty. Even so, the states remained reluctant to interpret their surrender on anything but the narrowest basis. This meant that for decades, the federal government was forced to leave to the individual states the necessary improvement in facilities for transportation and communication that would make it possible for so large and empty a territory to hold together and become a modern nation.

Despite the inefficiency inherent in each state's doing its own thing, toll roads and canals were built, particularly in the industrializing north. The most famous product of this age was the Erie Canal, which was opened in 1825 and which served to connect New York to the interior. New York, which till then had lagged behind Philadelphia as the national metropolis, suddenly forged forward to become and remain the largest city in the nation and the most remarkable city in the world (see Chapter 3).

Roads and canals have their limits, however. Men can only march so fast and horses can only gallop so fast on even the best road, while boats can only be poled along or pulled along just so quickly on even the best canal. Under those conditions, even roads and canals don't serve to knit a nation together when it is the size the United States grew to be.

To be sure, the ancient Roman Empire was larger than the United States of 1800, and it was held together by nothing more than roads on which horses galloped and armies marched and by seaways on which galleys were rowed by men and trading vessels sailed. The Roman Empire, however, had been built up by relatively slow accretion of (for the most part) already civilized areas and at its peak did not have to compete with nations that were more compact *and* more technologically advanced.

The earlier Persian Empire, on the other hand, though as large as the Roman Empire and held together in the same fashion, *did* have to compete with smaller states that were more technologically advanced. Down went Persia, therefore, before Alexander the Great of Macedon, in a campaign that is always treated as a semimiracle but was actually a foregone conclusion. Alexander was by no means David fighting Goliath; he was Hunter shooting Elephant.

The United States in those first decades of independence faced Europe in the position of the Persian Empire facing the Greek world. The situation was made worse (on a short-term basis) when President Jefferson carried through the Louisiana Purchase in 1803 and doubled American territory without doubling American ability to hold all that territory together against the strain of external pressure.

What saved us was, first, the insulating effect of three thousand miles of ocean between ourselves and Europe. Secondly, Great Britain, which alone among the European states could cross the ocean freely, was occupied with Napoleon. (Eventually, Great Britain was reluctantly forced into war with us in 1812 and, with both eyes on Napoleon, with only an occasional glance in our direction and on battlefields three thousand miles from home—she held us to a draw.)

But if circumstance saved us in our early, very vulnerable time, what was it that strengthened us (even as our area grew still larger till it was as large as all Europe) and kept us from falling apart under the terrific strain of the Civil War?

The answer lay in advances in technology, of course, and for the details there, see Chapter 5.

(*Note:* After this essay first appeared in print, in July 1976, I received a letter from Albert G. Hart of the Department of Economics of Columbia University, who said, "You really bat extraordinarily well"—which pleased me so much, since we all know I am not really an economist—but he went on to point out several places where I was off the beam.

The American colonies were by no means kept in complete economic subjection by Great Britain, he says. The colonists were encouraged to engage in shipbuilding, since they had enormous reserves of timber and the British had

65

not. Also he thinks I underrated the importance of water transport—the coastal trade, the rivers, and eventually the Great Lakes.

The Americans, he points out, developed the technology of interchangeable parts, independently and ahead of Great Britain, and had a thoroughly European system of education, which is something of considerable importance.

I have to give in. But even allowing for the colonies not being as dramatically in the soup as I made it appear, I still stand by my article, once we imagine it toned down a notch or two.)

## Five
# MOVING AHEAD

People sometimes get impatient with me because I keep insisting that nothing that has even taken place in human history can be properly understood without taking into account the effect of technological change. I had a friend, twenty years ago, who kept saying to me, "How do you explain the Crusades in terms of technological change?"

I knew what he meant. He thought it was all a matter of religious enthusiasm—of knights being caught up in a vision of the Holy Land under the heel of the foul Paynim.

I thought about it and, eventually, I said: "About the year 1000 the technological collapse in the West European provinces of the Roman Empire began to be alleviated. The invention of the moldboard plow meant that the dank soils of northwestern Europe could be efficiently turned. The invention of the horse collar and the horseshoe meant that the more efficient horse could replace the ox as a plow puller. All this meant that food production shot up and population increased.

"With the nobility increasing in number in particular (they got the best of the food) while the quantity of land did not, the average size of the fief decreased, and even so there were an increasing number of landless knights. These made the air of northwestern Europe horrid with their eternal fighting, and, by 1095, the Pope was only too glad to get rid of quantities of them by sending them off to the East to inflict themselves on the foul Paynim. Religion was the excuse, not the basic cause."

My friend couldn't bring himself to accept that but I wonder what he woud say today. Out in Lebanon (which was part of the Kingdom of Jerusalem that was set up in 1099 by the Crusaders) there is a civil war (as I write) between the Moslems and Christians, with the Christians badly outnumbered and sure to be defeated in the not-so-long run. This is precisely the situation that started a Crusade every time nine centuries ago. Even as recently as 1958, a similar but far less dangerous civil war in Lebanon had President Eisenhower sending in the Marines.

And now? No Christian nation says a word. Not a whisper. They're all looking off there in the middle distance.*

Why not? Is it because of the decline in religious fervor in the West? Partly, I suppose (and that is, to a great extent, because of the rise of science-based technology in recent centuries).

Is it because Western Christianity no longer presents a united front, thanks to the Protestant Reformation and the growth of secularism? Partly, I suppose (though both the Reformation and secularism would scarcely have been possible without the printing press).

But does anyone doubt that, for the most part, it's a case of religious sympathy being totally outweighed by the fears of an oil boycott, for perfectly obvious technological reasons?

Or let's take something that seems even more removed from technology than the Crusades do—the quality (such as it is) of my writing.

Reviewers are prone to talk about my "enthusiasm," my "liveliness," my "warmth." My excitement in my subject seems to stick out all over my writings and one would suppose that this is the result of my own particular ebullient and extroverted personality.

I can't deny that I *am* ebullient and extroverted, of course, but, nevertheless, technology is also required. The reason the ebullience has a chance to show up in my writing is that it doesn't evaporate in the process of converting thoughts into words. I think quickly in unspoken words and sentences and I need a device that will put them on paper as fast as they are created.

A quill pen wouldn't do it; nor would a steel pen, a fountain pen, or a ball-point pen. I can, and have, written articles and stories the size of this one or even longer with pen and ink, but it's a painful job and I can't keep it up very long at a time. If that were the only way I could write, the quality, I assure you, would be different—more

---

* This article was written in January 1976. Later that year, the Christians *were* rescued, but by the Syrian Moslems.

somber. Even an ordinary typewriter won't do, since I wear myself out, pounding those keys, after an hour or so.

No, what is needed is an electric typewriter, which requires only a delicate touch and on which I can stroke out my ninety words a minute through a whole, reasonably long working day, if I feel like it. Such a typewriter allows me to see what I write (dictating is like walking through a crowded thoroughfare with one's eyes closed) and it keeps up with me, so that I lose none of my excitement in the irrelevant processes of shaping a letter or pushing down a key by force.

So now let's go on analyzing American history in terms of technological advance—something I began in Chapter 4.

I ended Chapter 4 by pointing out that Frederick the Great had predicted that the United States was too large to hold together. And yet it *has* held together even under the explosive, disintegrating force of the most systematically resolute civil war ever fought.

How?

It was a matter of transportation and communication. Frederick felt that neither messages nor goods could travel from end to end of the new nation quickly enough, so that different parts would lose touch with each other and tend, in time, to go their own way.

It didn't occur to him that there might be fundamental changes in transportation and communication. Why should there be? There hadn't been for four thousand years. To be sure, the steam engine had been invented as a new source of power, but it was just beginning to show its potentialities at the time the United States had won recognition of its independence and Frederick missed that point.

Others did not. If the steam engine could turn a wheel in a textile factory, it could turn a wheel on the side of a boat, and if that wheel were equipped with paddles, the steam engine would become, so to speak, a mechanical and tireless galley slave who would row a ship against wind and current.

The basic concept was simple, and in 1785 John Fitch (born in Windsor, Connecticut, on January 21, 1743) had already thought of it. By 1790 he had a steamboat traveling up and down the Delaware River from Philadelphia to Trenton on a regular schedule.

Unfortunately, John Fitch's middle name was Bad Luck. Nothing had ever worked for him. He had had little schooling, a harsh father, and a nagging wife (whom he deserted). When he made some money with a gun factory in the Revolutionary War, he was paid in Continental currency that became worthless. The last part of the war he had spent as a British prisoner.

Even with his steamship running, then, after superhuman efforts to raise capital and to square the legal aspects with five different states, he found he couldn't inveigle passengers aboard. His financial backers drifted away and when a storm destroyed his ship in 1792, he was through.

He went to France to try again, but got there in 1793 at the most turbulent portion of the French Revolution, and could obtain no funds. He returned to the United States and died in Bardstown, Kentucky, on July 2, 1798, probably by suicide.

Do you think that ended his bad luck? Not at all! He invented the steamship, but how many people know that? Ask anyone, and you'll be told that Robert Fulton did.

Fulton was born in Little Britain, Pennsylvania (a town that is now called Fulton), on November 14, 1765. About the time Fitch died, Fulton, who had gone to Great Britain after the Revolution, began to think about powered ships.

In 1797 he went to France and spent years trying to devise a workable submarine. His most nearly successful one was built in 1801 and he named it *Nautilus*. The name at least was a triumph. In 1870 the French writer Jules Verne wrote *Twenty Thousand Leagues Under the Sea* and named Captain Nemo's submarine *Nautilus*, after Fulton's craft. Then, in 1955, the United States launched the first nuclear-powered submarine, and named it *Nautilus*, after Nemo's craft.

Fulton also worked on surface vessels and tried to run a steamboat up and down the Seine River. It didn't work, and although Great Britain and France (who were in the first years of what was to prove a twenty-year war) were sensible of the war application of steam propulsion, neither nation felt like investing much in crazy schemes.

In 1806 Fulton returned to the United States and continued his experiments on the Hudson River. He managed to get the financial backing of Robert R. Livingston (one of the five-man committee who had once been charged with writing the Declaration of Independence, and a man who had served as the United States Minister to France when Fulton was there). Fulton built a steamship he called *Clermont* and on August 7, 1807, she began chugging up the Hudson River. She reached Albany in thirty-two hours maintaining an average speed of 8 kilometers (5 miles) per hour.

Though Fulton did not build the first working steamship, he built the first one to make a profit and I suppose that counts. He died of pneumonia on February 24, 1815, contracted after working on the open deck of a steamboat under construction during bad weather, but by that time there was a fleet of steamboats operating under his direction.

What the steamboat did for the United States was to make the great rivers of the nation highways in both directions, *up*stream as well as down. By the 1850s the steamboat on the Mississippi River was living through a golden age, forever enshrined in Mark Twain's *Life on the Mississippi.*

From earliest times, the sea was easier to cross than the land was. The sea was level and navigable in all directions; the land is hilly, swampy, rocky, sandy, and, in general, navigable with only the greatest difficulty on anything but legs except where decent roads exist, and these, until the twentieth century, were so few as to be almost nonexistent.

The turning point came when the steam engine was used to turn wheels on a locomotive ("move from place to place") that could, in turn, pull a train† of passenger or freight cars behind it. The amount of energy that would have to have been used to drag the wheels of all those cars over uneven, rocky, muddy ground would have been quite unthinkable, so the trick was to lay down a pair of parallel rails (first of wood, eventually of steel) on which the wheels could turn, unimpeded as a ship moving over the sea.

The inventor of the steam locomotive was the Englishman, Richard Trevithick, born near Illogan, Cornwall, on April 13, 1771. As early as 1796, he was designing steam locomotives and was the first to demonstrate that smooth

† The word can be applied to any series of like objects in single file.

metal wheels could find enough traction on smooth metal rails to allow motion, thanks to the weight of the locomotive pressing one against the other.

In 1801 Trevithick had locomotives in operation, but, like Fitch, he was plagued with misfortune. Though his locomotives worked, he had to face insufficient steam, too much fire, broken axles, public hostility, and so on. In the end he gave up and went to South America to sell steam engines.

As in the case of Fitch, the credit for Trevithick's invention went elsewhere. Unlike Fitch, Trevithick lived to see that.

The inventor who got the credit was George Stephenson, who was born in Wylam, Northumberland, on June 9, 1781. He had the advantage of having a father who worked with steam engines and who introduced him to the field. He had the disadvantage of being uneducated and illiterate. In his late teens he attended night school to learn to read in order that he might be able to go through the works of James Watt.

He began to build locomotives, and in 1825 one of his locomotives pulled thirty-eight small cars along rails at speeds of from 20 to 25 kilometers (12 to 16 miles) per hour. It was the first *practical* steam locomotive built, and by 1830 Stephenson and his coworkers had eight engines at work on a railroad between Liverpool and Manchester. For the first time in the history of the world, land transportation at a rate faster than that of a galloping horse became possible.

(Poor Trevithick was still in South American and, with his usual bad luck, found himself caught up in the colonial revolutions against Spain and forced to fight on the side of the rebels. The only way he could return to England was, ironically enough, by borrowing money from Stephenson's son, who happened to be in South America at the time and whose money came from the dividends of his father's successful railroad. Trevithick died in Dartford, Kent, in poverty, on April 22, 1833.)

The United States was neck-and-neck with Great Britain as far as the railroad was concerned. In 1825 one John Stevens built the first locomotive in the United States to run on rails—on a half-mile track near his home in Hoboken, New Jersey.

In 1827, the Baltimore and Ohio Railroad was chartered. On July 4, 1828, the fifty-second anniversary of the Declaration of Independence, work on the first passenger and freight railroad in the United States began in Baltimore. The ground was broken by Charles Carroll, who at that time, at the age of ninety-two, was the last surviving signer of the Declaration of Independence.‡ On May 24, 1830, the first thirteen miles of track opened.

More than any other nation in the world, the United States threw itself into an orgy of railroad building. Within ten years, its lines of railroad track numbered 4,500 kilometers (2,800 miles) and within thirty years, 48,000 kilometers (30,000 miles).

Through all the history of the world, transportation and communication were almost synonymous. In general, a message came only with a messenger who had to cover ground on his feet, on a horse, on a ship, or, for that matter, on a railroad. The only messages that could arrive ahead of a messenger were those that were sent by sight or sound—semaphore signals, reflection flashes, smoke signals, tom-toms, and so on. These were all limited in range. The turning point came with the use of the electric current.

The Italian Alessandro Volta invented the chemical battery in 1800, and an electric current was reliably produced for the first time. The Dane Hans Christian Oersted discovered electromagnetism in 1820, and, immediately afterward, the Frenchman André Marie Ampère worked out the theory of electric currents. The Englishman Michael Faraday introduced the electric generator in 1831, and made the current cheap enough for routine and massive use. The American Joseph Henry invented the electromagnet with insulated wire in 1829 and, in 1831, the electric relay and the electric motor—all of which made the current open for use in a versatile way.

The first striking application of the current came through the work of an artist, Samuel Finley Breese Morse, born in Charlestown, Massachusetts, on April 27, 1791.

Morse is not, to me, a very attractive person. He did not feel any particular bonds of patriotism to the United States and lived in Great Britain at ease throughout the

‡ Carroll died on November 14, 1832, aged ninety-five years, two months.

73

War of 1812. When he returned to the United States and entered politics, it was as a member of the Native American party (a group of bigoted anti-Catholic and anti-immigrant people, usually and appropriately called the "Know-Nothings"). During the Civil War, he was strongly pro-South, since he was an anti-Black who believed slavery a good thing.

During the 1830s, Morse caught the fever of electrical experimentation from the American chemist Charles Thomas Jackson, a fellow passenger on an ocean voyage. (Jackson was an eccentric scientist of considerable brilliance who half-discovered a number of things, engaged in endless quarrels over priority, and died insane.)

Morse knew little about electricity but met Joseph Henry by chance. Henry, a person of warm benevolence, helped Morse without stint, answering all his questions and explaining the workings of the electric relay. Morse, a person of icy selfishness, absorbed everything and in later legal battles over priority tried to pretend he had obtained nothing from Henry.

Both Henry and the British physicist Charles Wheatstone had built working systems of what came to be called the "telegraph," but Morse added something of considerable importance, a system of clicks spaced at long and short intervals that could serve as a "code" named for him for sending telegraphic messages. He also contributed a kind of remorseless promotorship that succeeded in raising money from unlikely sources.

He obtained a patent for his telegraph system in 1840, then managed to persuade and bully a most reluctant U. S. Congress into appropriating $30,000 for his use in 1843—by a margin of six votes—to build a telegraph line over the forty-mile stretch from Baltimore to Washington. It was completed in 1844 and Morse's first message, in Morse code, was "What hath God wrought?"—a quotation from the Bible (Numbers 23:23). For the first time a messengerless message could, in principle, be made to go *any* distance all but instantaneously.

Before the year was out, the telegraph was being used by the nation's reporters to announce the details of the Democratic presidential nominating convention. Before five years had passed, telegraphic communication was established between New York and Chicago, and by the time of the Civil War, the telegraph lines crisscrossed the nation.

By the 1860s, then, the United States was knit together by land, by sea, and by wire, and it could not fall apart by centrifugal force alone, simply because one part was out of touch with the other. Frederick the Great was wrong; he had not taken technological advance into account.

But the United States *did* threaten to fall apart, not through mere incoherence, but over deep-seated and violent differences between the northern and southern tiers of states—differences that resulted in a dreadful war. How did the matter of technology prevent collapse in this case?

The northern states, which fought for the Union, had a half-century history of growing industrialization. They produced iron and steel in quantity so that rails could easily be laid down and locomotives easily built. Northern industrialists, anxious to ship out their goods and bring in their raw materials, pushed for railroad lines across the various states and the federal government was willing to co-operate in this.

The result was that, by 1861, two thirds of the miles of railroad track in the United States existed in the northern states, and those lines were a well-integrated network.

The southern states, however, adhered to their belief in the virtues of Jeffersonian ruralism, and the large plantations tended to be much more self-sufficient than the units of the northern social system. There was a correspondingly weaker push toward railroads in the South, which, in any case, found railroads to be economically disadvantageous since all finished materials and trained engineers had to come from the North or from Great Britain.

Furthermore, since the South was strong on States' rights as a way of protecting itself from the more populous North, which was increasingly dominating the Union, each southern state built its railroads as it saw fit without concerning itself overmuch with its neighbors. The result was that the thinner railroad network in the South was not well-integrated and did not even utilize mileage properly.

The Civil War, fought by armies of hundreds of thousands over a battlefield that stretched over thousands of square miles, presented enormous transport and supply problems for both sides. Either a mass army must have its food and clothing brought to it by mass transportation of some sort or it must live off the surrounding countryside.

Since the war was fought on southern territory, the Union armies from the North might, on occasion, choose to live off the country as a means of weakening the morale of the enemy. Both Sherman in Georgia and Sheridan in the Shenandoah Valley did this. Except as a deliberate army policy of *Schrecklichkeit*, however, the North didn't have to resort to this. Their railroads worked and their armies were well supplied (except where dishonest contractors and dishonest politicians connived to sell them junk and garbage).

The southern armies, however, could not very well deplete their own countryside without destroying their own cause. But if they didn't, they were in trouble, for their railroad network was inadequate for the job, and what's more, as southern railroad equipment wore out or broke, there was almost no way of getting replacements.

The result was that the Confederate armies were always underfed, underclothed, and underarmed. They performed prodigies of valor, but to what end? The Union armies simply hung on until the factory workers from the northern cities learned to fight as well as the farm boys and horsemen from the southern farms. Once that happened, the end was in sight for the South.

What's more, while the South slowly withered and died behind the northern blockade, the North actually grew economically stronger as the war progressed—thanks to technology.

In 1834 Cyrus Hall McCormick, born in Rockbridge County, Virginia, on February 15, 1809, patented a horse-drawn mechanical reaper that rendered unnecessary all the bending and cutting that had always been involved in the process and let one man do the work of many.

Although McCormick was a southerner, it was not the South that demanded the machine. There were slaves to do the work there cheaply enough, and to get the machines might raise the prospect of having to support slaves in idleness.

McCormick therefore set up his factory in Chicago, for the Midwest had far more acres than farm workers and something labor-saving was needed. Within a year he had sold eight hundred reapers, and in the 1850s, he was selling four thousand a year.

The mechanization of agriculture was well begun and the Midwest began to produce grain at an unprecedented rate. During the years of the Civil War the North could sell quantities of grain to a hungry Europe in return for everything it needed to keep its industries humming—while southern cotton and tobacco rotted in the fields and warehouses behind the northern blockade.

Even the loss of manpower was not serious in the North (economically, that is, for no one can measure the personal suffering produced by the bitter bloodshed among the soldiers and those who loved them).

All through the decades preceding the Civil War, immigrants from Europe had gravitated to the North, where there were factories and farms and railroads to employ them and prosperous individuals who needed servants (who were free and who could quit when something better turned up). Few immigrants, on the other hand, tended to go to the South, where it was difficult to find work in competition with slaves, where unskilled labor had, in any case, the cachet of slavery about it, and where the mystique of family and breeding limited upward mobility sharply.

During the Civil War immigration to the North actually increased, as industry and agriculture operated at forced draft, while to the South it stopped altogether. The immigrants came to the North in such oversupply that they overflowed into the army. One third of the soldiers in the northern armies were foreign-born.

When Grant caught Lee in the climactic battles of 1864 in Virginia, Grant could easily afford to lose two men to every one that Lee lost. Grant could rely on endless reinforcements, while Lee's losses were irreplaceable. Grant understood this and attacked steadily and relentlessly. He was called "the butcher," but he won the war.

And when the war was over, the United States had so profited from it technologically that in wealth and in strength it began to move ahead of the nations of Europe, even ahead of the proud empire of Great Britain.

This was by no means noticed at the time, however. Through habit, the American was looked upon by the European as a frontiersman without culture; as a kind of crude barbarian who did have a certain knack for adding up profits. Otherwise, he was sneered at and never taken seriously.

But you know, though all through history knights have sneered at merchants, the fact is that in the long run the merchants win and the knights lose. The Dutch merchants beat the Spanish knights, and the British beat Napoleon who thought "perfidious Albion" was only a nation of shopkeepers.

It was now the turn of the United States.

It is not surprising that the truth was more nearly plain to science fiction writers than to diplomats and generals. In 1865, when Jules Verne published *From the Earth to the Moon*, about the first astronauts being shot to the Moon by a giant cannon—to whom did he attribute the feat? To Americans, of course. They made up the nation, he saw (one century in advance of fruition), that would reach the Moon.

And when did the rest of the world see it? —Well, that's for Chapter 6.

*Six*
# TO THE TOP

In the first half of February 1976 I was on the *Queen Elizabeth 2*, taking a Caribbean cruise with my wife, Janet.*

We had a table for two in one of the dining rooms (where, it seems to me as I think back upon it, we spent most of our time); and to our left was another table for two, at which there sat a very pleasant Austrian and his equally pleasant young daughter. This was delightful, because it gave me a chance to practice my well-known suave approach to young women *in German*.

The Austrian, who spoke English, talked constantly of the delights of his native province of Carinthia (which he called "Kärnten," for some reason, but I was too polite to correct him) and of Vienna. He spoke very convincingly, too, so that while I did not feel impelled to go to Carinthia, since I do not like to travel, I caught myself wishing someone would bring Carinthia to New York.

In particular, whenever some Continental item was on the menu (which was, of course, fabulous, to the delight and distress of my waistline) he would order it, taste it, give his head a good-natured shake, and say, "We do it better in Austria." It got so that I could tell exactly when he was going to say it and would say it with him and we would both laugh.

As it happened, he was a great traveler and he was shocked at discovering that I had enough money to travel wherever and whenever I pleased, yet chose not to do so. He undertook to lure me into it by describing the wonders he had seen, and he waxed especially poetic over the Grand Canyon. When he ran out of English, he switched to German and continued.

* No, I was *not* taking a vacation. I gave two talks on the ship and a third on the island of Barbados, and I wrote two stories longhand.

"It sounds as though you liked the Grand Canyon," I said.

"Liked it?" he said. "It was a magnificent, an incredible sight."

And without as much as a facial twitch, I said gravely, "But you do it better in Austria, right?"

"Well, no," he said. *But he hesitated.*

There's nothing wrong with a bit of local pride, however. I've got it myself. I'm actually very fond of the United States and that's why, when I started out to write an essay on the climb of the United States to world leadership in technology, I found myself lingering over the subject and stretching it out over three essays.

In Chapter 5 I carried the United States through the Civil War and pointed out that by then, the nation was well on the road toward technological leadership. It was still a distant second to the established leader, Great Britain, in coal and iron production, but in all respects it was moving up rapidly.

The question is, however, when did people actually come to realize that the United States was going to be the new leader? In a way, there was already a kind of mute awareness of the fact, for Europeans were emigrating to the United States by the millions. Between 1870 and 1890, a hundred thousand a year were coming into the United States, some even from the British Isles.

In another way, some of the more parochial types, in Great Britain particularly, never got rid of the comic stereotype of the American backwoodsman. Even as late as the 1930s Agatha Christie, in her mysteries,† would frequently introduce American characters who would always have a first name like Hiram, who would speak in a nasal twang, who would start every other sentence with "I reckon," and who, on the whole, acted as though it were 1840. I used to watch for some indication that one of her Americans chewed tobacco and brought his Black slave with him.

---

† These, in my opinion, are the best ever written and I deliberately and consciously imitate them when writing my own mysteries—though I introduce my own improvements, of course, as you will see if you read my recently published *Murder at the ABA* (Doubleday, 1976).

Still, there might be a turning point; some moment when you can say: "It was at this very time that American technological leadership had to be taken seriously." I have a candidate for that turning point. It has a name, and it has a year.

The name, first. It is Thomas Alva Edison.

Edison was born in Milan, Ohio, on February 11, 1847, and was the son of a Canadian immigrant who was, in turn, descended from an American Tory who had fled to Canada after the Revolutionary War. Edison's life is the classic tale, so beloved by Americans, of the self-made man—of the poor boy who, without schooling or influence, made his way to fame and fortune through hard work and intelligence.

He was a puzzling boy from the start. His curious way of asking questions was taken as an annoying peculiarity by the neighbors. When he made little progress at school, his mother inquired and was told by the schoolteacher that the boy was "addled." His mother, furious, took him out of school. She was in any case concerned for his delicate health and, being a schoolteacher by profession herself, could easily supervise his primary education.

Edison turned to books as a supplement. His unusual mind then began to show itself, for he remembered almost everything he read, and he read almost as quickly as he could turn the pages. He devoured nearly everything, though he found Newton's *Principia Mathematica* too much for him—but then he was only twelve years old at the time.

When he began to read books on science, he wanted to set up a chemical laboratory of his own. In order to get money for chemicals and equipment, he went to work. At the age of twelve he got a job as a newsboy on a train between Port Huron and Detroit, Michigan. (During the stop at Detroit, he spent his time in the library.)

Selling newspapers wasn't enough for Edison. He bought second-hand printing equipment and began to publish a weekly newspaper of his own, the first newspaper ever to be printed on a train. With his earnings, he set up a chemical laboratory in the baggage car. Unfortunately, a chemical fire started at one time and he and his equipment were thrown off the train.

In 1862 young Edison, in true Horatio Alger fashion, saw a small boy on the train tracks and, at the risk of his life, snatched him from the path of an oncoming loco-motive. The grateful father, who had no money with which to reward the young man, offered to teach Edison telegraphy in return. Edison was eager to learn and quickly became the best and fastest telegrapher in the United States. He earned enough money at his new pro-fession to buy a collection of the writings of Faraday, which solidified his interest in electrical technology.

In 1868 Edison went to Boston as a telegrapher and that year patented his first invention, a device to record votes mechanically. He thought it would speed matters in Congress and that it would be welcomed. A congressman told him, however, that there was no desire to speed pro-ceedings and that sometimes a slow vote was a political necessity. After that, Edison decided never to invent any-thing unless he was sure it was needed.

In 1869 he went to New York City to find employment. While he was in a broker's office, waiting to be inter-viewed, a telegraph machine broke down. Everyone pres-ent was helpless, but Edison's quick eye saw the component that was out of place. He offered to fix it and did so; and was promptly offered a better job than he had expected to get.

In a few months he decided to become a professional inventor, beginning with a stock ticker he had devised during his stay in Wall Street. He planned to offer it to the president of a large Wall Street firm and to ask five thousand dollars for it. As he waited for the interview, however, five thousand seemed to be more and more astro-nomical, and when it was time to talk, he lacked the courage to put the request into words.

"What would you be willing to pay for it?" he quavered.

The Wall Street man said, tentatively, "Forty thousand dollars?"

Edison, still only twenty-three, was in business. He founded the first firm of consulting engineers the world had seen, and for the next six years he worked in Newark, New Jersey, turning out inventions such as wax paper and the mimeograph, to say nothing of introducing im-portant improvements in telegraphy. He worked about twenty hours a day, sleeping in catnaps, and developed a group of capable assistants. Somehow he found time to get married.

In 1876 Edison set up a laboratory in Menlo Park, New Jersey. It was to be an "invention factory," and eventually he had as many as eighty competent scientists working for him. It was the beginning of the modern notion of the "research team."

He hoped to be able to produce a new invention every ten days. He didn't fall short of that, for before he died he had patented nearly 1,300 inventions, a record no other inventor has ever matched. In one four-year stretch, he obtained three hundred patents, or one every five days. He was called "The Wizard of Menlo Park" and in his lifetime it was already estimated that his inventions were worth $25 billion (1930-type dollars, too) to humanity.

In Menlo Park he invented the phonograph, which was his own favorite invention.

Then came 1878. If the recognition of the entry of the United States into technological leadership bore the name "Edison," it also bore the date "1878." To explain that, let's go back in time.

Before human beings began to fiddle with the Universe, there were three types of light on Earth:

1. There was light from the sky: the Sun, the Moon, the planets, the stars, the lightning.

2. There was light from living creatures, such as fireflies.

3. There was light from spontaneous fires, usually caused when a bolt of lightning struck a tree.

The Sun, however, is absent from the sky an average of twelve hours a day. The Moon is a feeble substitute and is, on the average, absent for half of each night. The other heavenly bodies, the lightning, the fireflies are all insignificant. Forests fires are an absolute danger.

If early hominid slept eight hours a day as we do, he was immobilized, on the average, for one third of each night, lying in the dark and waiting for the dawn.

Hominids more primitive than *Homo sapiens* learned, however, to tame fire and, eventually, to produce it on demand. In addition to supplying heat and making various new technological advances possible (metallurgy, for instance), fire also made it feasible for human beings to be active an additional four hours a day on the average, lengthening the effective length of life by about 17 per cent.

Lightning has remained a vital need of mankind ever since those early prehistoric days, and through all the hundreds of thousands of years—right down to one century ago—human beings produced their needed light by combustion—by burning something.

The best fuel for lighting would be something that would burn slowly and produce as much light, along with the heat, as possible. Ordinary wood is not ideal for the purpose. Resinous wood is much better and makes good torches.

Animal fat is less common than wood but, ounce for ounce, produces more light more conveniently. From solid fat, candles could be made, with wicks running their full length. A wick can also be floated in liquid oil kept in a container (a "lamp," from a Greek word meaning "to give light").

All these sources of light—bonfires, torches, candles, lamps—are of prehistoric origin, and nothing essentially new was added through all of history right down to the nineteenth century.

With the nineteenth century the pace of change quickened. Nothing much could be done with wood fires; and coal fires, which now became common, though an improvement as far as heat was concerned, were poorer as light sources. The story was different with fats and oils.

In 1835 the French chemist Michel Eugène Chevreul, who had isolated fatty acids from natural fats and oils, patented a process whereby candles could be made from those fatty acids. Such candles were harder than earlier candles, burned more slowly and brightly, and gave off considerably less odor.

As for liquid fuels, whale oil turned out to be particularly useful in lamps, thus contributing an impetus toward the remorseless slaughter of these largely inoffensive sea creatures. That was later supplemented by kerosene, which was obtained from petroleum.

The great advance in lighting of the early nineteenth century was, however, the introduction of gas lighting. Gases had the potential of burning more clearly and less smokily than either solids or liquids. They could be led by pipes to the point desired from some central storage place and the amount of light produced could be more easily regulated than was true for liquid or solid fuels.

The first public use of gas for lighting was in Paris in 1801. It was arranged by a French chemist, Philippe Lebon, who got the gas for the purpose by heating wood in the absence of air ("destructive distillation"). He had been experimenting with gas lighting since 1797, worked out much of the engineering requirements, and foresaw all the applications thereof. France, however, was in the midst of the Napoleonic wars at the time and Lebon himself died in 1804, so the lead in gas lighting passed to Great Britain.

There, the Scottish inventor William Murdock was also working on gas lighting. He got his inflammable gas from the destructive distillation of coal. He put up his first public display of gas lighting in London in 1802 to celebrate the temporary Peace of Amiens with Napoleon. In 1803 he was routinely lighting his main factory with gas jets, and in 1807 some London streets began to use gas lighting.

By 1825 gas lighting had grown common in the public buildings in London and in the factories and shops as well, but for years the flames were sooty and smelly. It wasn't till methods for introducing air into the gas tube just before burning were worked out that the flame became clean and nonodorous. This took place about 1840. (In 1855 the German chemist Robert Wilhelm Bunsen devised a simple version of such a gas jet for laboratory use, and this "Bunsen burner" has been enormously useful in the chemistry laboratory ever since.)

By the 1870s, then, gas lighting was the chosen method of illuminating the streets and homes of the cities and towns of the more advanced nations.

Like all other methods of lighting, however, from the original wood fire, the gas jet involved an open flame. Indeed, as long as light was obtained from combustion, an open flame would seem a necessity, since oxygen from the air had to reach the burning fuel and the carbon dioxide produced had to leave.

The open flame had been with mankind for half a million years, then, and it was dangerous. Who can count how many times that barely controlled flame went out of control, burning down wooden houses and wooden cities and destroying flesh-and-blood human beings in agony.

What's more, open flames were generally dim (by modern standards) and invariably flickery. Reading by any sort of open flame or doing any fine work by it must surely have been very tiring on the eyes in view of the constant shifting and bellying of shadows.

But how can you have light without an open flame? How can you have light without combustion?

The first indication that such a thing was possible arose from an observation of sparks from static electricity devices.‡ Making use of batteries to produce a constant electric current, one could set up a permanent electric spark between two carbon electrodes. The electrical engineer W. E. Staite experimented for years with such "arc lights" and, beginning in 1846, gave impressive public demonstrations of their use.

The arc light was far brighter than an ordinary flame. To be sure, it was just as hot and just as fire-provoking as a flame, but it didn't require a steadily renewed current of air to maintain itself or to carry away wastes, so it could be enclosed in a glass container. However, the spark flickered even more than a flame did and it was hard to regulate.

One way of getting the electricity-produced light to remain steady was to run an electric current through a wire of very-high-melting metal and let the wire heat to incandescence. The wire would not flicker but would remain in place, and so would the light. Unfortunately, at those high temperatures, the metal will burn. Even the high-melting and inert metal platinum will slowly combine with oxygen and break, so that such an "incandescent light" would only last a short time.

The obvious trick, then, was to enclose the glowing wire in an evacuated glass container. Then there would be no oxygen for the metal of the wire to combine with. It is easy, however, to *talk* about an evacuated glass container; producing one with a good enough vacuum is a lot harder.

From 1820 on, inventors (mostly in Great Britain) tried to produce what we would now call an electric light bulb. The most nearly successful experimenter in this direction was the nineteenth-century English physicist Joseph Wilson Swan. He was the first to see clearly that even if he could produce a practical light bulb with a platinum filament, it would end up being too expensive for mass use.

‡ See "The Fateful Lightning," in *The Stars in Their Courses* (Doubleday, 1971).

It occurred to him that carbon had as high a melting point as platinum and might substitute for it.

Of course, carbon isn't a metal and can't be drawn into wires as metal can. In 1848, however, Swan began to use thin strips of carbonized paper within an evacuated glass bulb. For nearly thirty years he kept fiddling with this and improving his design, but always he was defeated by one thing—the vacuum in the bulb was never good enough, and after glowing briefly, the carbon filament he used burned, broke, and went dark. (So did platinum filaments.)

By 1878, then, inventors had been working on the electric light bulb for over half a century and had gotten nowhere. And then, in that year, Thomas Alva Edison, the Wizard of Menlo Park, announced that *he* would try.

*And on that bare announcement*, illuminating-gas stocks tumbled on the stock markets of New York—and of London, too! Faith in the young man (he was only thirty-one) was absolute!

It's my feeling that the moment those stocks fell, there was a clear-cut indication that the investing community of Great Britain was taking American technology seriously, and that was the moment you might suppose that the clear suspicion arose that world leadership in technology was crossing the Atlantic.

Nor did Edison disappoint the world. By that time the art of preparing an evacuated bulb had reached the point where an electric light was possible and it remained to find the proper filament. Apparently, Edison was not thoroughly aware of Swan's work, for it took him a year of experimentation, and fifty thousand dollars, to find that platinum wires wouldn't do, and to turn to a scorched cotton thread.

On October 21, 1879, Edison set up a bulb with such a carbon filament that burned for forty continuous hours. The electric light was at last a reality and it received U.S. patent number 222,898. On the next New Year's Eve the main street of Menlo Park was illuminated by electricity in a public demonstration which three thousand people (mostly from New York City) watched.

In order to make the electric light commercial, Edison had to develop an electric generating system that would supply electricity when needed and in varying amounts, as lights were switched on and off. This required more ingenuity by far than the electric light itself, but by 1881

Edison had built such a generating station and within a year he was supplying about four hundred outlets divided among eighty-five customers.

Meanwhile, Swan, in Britain, had independently produced workable electric light bulbs and the House of Commons was lit by electricity in 1881. Edison and Swan settled differences among themselves and formed a joint company in Great Britain in 1883.

I don't have to emphasize what electric lighting has done for the problem of illumination in the last century and how unendurable it would be to go back to the open flame. For one thing, consider how, despite the possibility of defective wiring, the danger of fire has been enormously decreased by the use of enclosed sources of light and by the elimination of open flames.

Of course, accidental fires continue because the open flame has not been entirely abolished. There are still open flames in gas stoves, furnaces, and internal combustion engines. Most of all, there is that near-equivalent of an open flame that hundreds of millions of people the world over carry about with them at all times—the smoldering cigarette end and its accompaniment, the smoldering match. *There* are the true villains!

Of course, you might argue that Edison's announcement and the accompanying stock tumble didn't indicate a recognition of American technological mastery at all—just of Edison's.

That, however, is not so. Edison just happened to be the best and most glamorous example of what was going on in the United States, but he was by no means an isolated example. He was just the leader of a large pack, and in the glare of his genius, American technology shone brilliantly from sea to shining sea.

The virtual explosion of technology that was going on in the United States in the last half of the nineteenth century was fed by the nation's free immigration policy, too, for from all over Europe brains as well as hands were pouring into the country.

It was a Swedish immigrant, John Ericsson, who built the U. S. Navy ironclad *Monitor* in 1861 and made every other warship in the world obsolete. It was a Scottish immigrant, Alexander Graham Bell, who invented the telephone in 1876. The German immigrant Charles Proteus Steinmetz and the Croatian immigrant Nikola Tesla advanced electrical theory and practice to an even greater extent than Edison did.

Then came 1898, which saw a demonstration of the effect of American technology in a way that left nothing to the imagination.

In that year the United States went to war with Spain. It was a manufactured war and Spain wasn't much of an enemy. Even so, the American Army was so small and so miserably mishandled that had the enemy been less incredibly inept than the Spanish forces in Cuba were, the United States would have ended up terribly embarrassed.

At sea, things were different. The American Navy was small compared to the British Navy, but is was newly built and its ships were as technologically advanced as the Americans could make them. The Assistant Secretary of the Navy, Theodore Roosevelt, had, in the absence of his superior, prepared for war by ordering six warships under Commodore George Dewey to Hong Kong where they would be ready to swoop on the Spanish fleet in the Spanish-owned Philippine Islands.

The war began on April 24, 1898, and as soon as the news of that reached Dewey, he sailed for Manila with his six ships. Waiting for him there were ten Spanish ships and Spanish shore batteries, and the British in Hong Kong were sure he was sailing to his destruction.

At daybreak on May 1, 1898, the Battle of Manila Bay began and in seven hours every Spanish ship had been sunk or beached and 381 Spaniards had been killed. No American ship suffered any significant damage, no American was killed, and only eight American sailors received minor wounds.

Meanwhile, in Cuba, another Spanish fleet was bottled up by another American fleet. On July 3 the Spanish fleet tried to make a run for it and the American ships pounced. In four hours every Spanish ship was destroyed, with the loss of 474 Spaniards killed and wounded and 1,750 taken prisoner. No American ships were significantly damaged, one American was killed, one wounded.

That the United States had won the war was not too surprising, but those naval victories were to the last degree impressive. Two naval battles had been fought, more or less simultaneously, on opposite sides of the world and both been ridiculously lopsided.

This could not be put down to the fact that the Spaniards could not fight, for military history has shown that Spaniards have consistently fought like demons under all conditions. Nor did Spain lack a naval tradition. For four centuries they had had a significant navy.

No, it was just this: in the last half of the nineteenth century, the art of naval construction had undergone enormous technological strides. No nation that was not technologically advanced could any longer fight a sea battle against one that *was* technologically advanced and inflict as much as a scratch on its enemy.

The United States had now demonstrated even to the very dense minds of the military establishments of the world that it was technologically advanced, and with the Spanish-American War it became a member of that dangerous group of nations known as "the Great Powers."

It was Great Britain's action that was most significant. Even as the United States was fighting Spain, Great Britain was preparing to use its unpracticed army to fight the Boers and was soon being handed humiliating defeats. It took Great Britain three years to win that war and in the process she came to realize that the whole world sympathized with the Boers.

In particular, Germany made no secret of its delight at British troubles and it was beginning to build an advanced navy of its own. Combine this with Germany's army—the best in the world—and it was clear that Great Britain's domination of the world was in jeopardy.

But throughout the nineteenth century Great Britain and the United States had been "traditional enemies" and no decade had passed without a war crisis between them. Great Britain now saw that she dared not let the German and American navies combine against her. From 1898 on, therefore, Great Britain never again allowed herself to be annoyed by the United States. Whatever the United States did, Great Britain smiled and nodded.

The result was that the two nations were enemies no longer. Throughout the twentieth century Great Britain and the United States fought together, through hot wars and cold, against Germany, Japan, and the Soviet Union.

You can analyze world events in terms of politics, ideologies, and international diplomacies all you want; my own feeling remains that it all boils down to technology.

# IV

---

# OUR PLANET

# Seven
# OF ICE AND MEN

Breathes there a man with soul so dead as never to suspect that there is a conspiracy against him on the part of the Universe?

As an example— I am frequently on the road to some speaking engagement or other. Since I don't fly, I get there by automobile. I am *convinced* that the incidence of rain on days when I drive is far higher than the incidence of rain generally.

I get a kind of dour satisfaction when I start out on a beautifully sunny day, with the weather bureau breathlessly predicting prolonged droughts, and then see the rain clouds gathering and the raindrops beginning to fall. It gives me a warm feeling to know that the countryside will get the welcome rain only because of me and my good old automobile.

Here's another case. I bought a house in Newton, Massachusetts, and moved in on March 12, 1956. For the first time in my life I was a landowner. It was a pleasant, middle-sized house, with a two-car garage beneath, and a nice, wide, deep driveway. No longer would I have to park my car at the curb.

On March 16, 1956, it started snowing. By the morning of March 17 there were three feet of snow in the driveway. I had never shoveled snow much in my life (one of the advantages of being a perennial tenant) but I had bought a snow shovel as one of the appurtenances of landed-gentryhood (I had also bought a lawn mower). I now took snow shovel in hand and got to work with rapidly diminishing enthusiasm.

For three days I sweated and shoveled and grunted and puffed, on and off, until my driveway was finally clear. By the morning of March 20 I could view a clear driveway again, set between icy mountains.

On March 20 we had a second storm and *four* feet of snow drifted into the driveway. It's a painful memory which I will not further elaborate upon at this time, but will someone kindly tell me why the worst double snow-storm in the history of the Boston weather bureau had to come during the very first week in which I owned a garage and driveway?

But there is a silver lining. That work represented my personal experience with Ice Ages and now makes it possible for me to write about them, and their effect on human beings, with a feeling of inner authority. I will, however, write about Ice Ages in my own inimitable fashion—very roundaboutly.

Imagine the Earth revolving about the Sun. The curve of its orbit is planar; that is, you can imagine an infinitely thin plane passing through the center of the Earth and the center of the Sun, and the Earth, as it moves around the Sun, will remain in that plane at all times.

If the Earth's axis of rotation were exactly perpendicular to the orbital plane, then the half of the Earth's globe which would be in sunlight would, at all times, be bounded at the north by the North Pole and at the south by the South Pole. As the Earth rotated on its axis and revolved about the Sun, that would not change.

If this were the case, a person standing at either the North Pole or the South Pole, on a stretch of ground per-fectly flat all the way to the horizon in every direction, would see the Sun forever at the horizon* and moving steadily about the horizon, east to west, and completing a circle every twenty-four hours.

In actual fact, though, the axis is tilted to the orbital plane by an angle of 23.44229 degrees, and this spoils that pretty picture.

Suppose we imagine the Earth to be located in its orbit in a place where the northern part of the axis is tipped directly toward the Sun (see Figure 1). The entire North Frigid Zone—all the Earth's surface within 23.44229 de-grees of the North Pole—is exposed to sunlight in that case.

---

* The effect of atmospheric refraction would actually lift it just above the horizon.

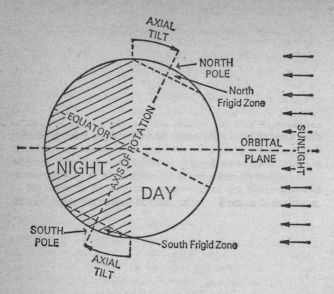

*Figure 1 — The Tilt of the Axis*

To any observer within the North Frigid Zone, under those circumstances, the Sun will be seen to circle the sky without setting. At the North Pole, the Sun will make a level circle, 23.44229 degrees above the horizon (if we ignore the effect of atmospheric refraction). At a distance from the Noth Pole, the Sun will make a tilted circle, reaching the highest point at noon and the lowest at midnight. At a distance of 23.44229 degrees from the North Pole, the Sun will touch the horizon at midnight.

The South Frigid Zone, on the other hand, under the same circumstances, will be entirely in darkness, and the Sun will not rise at all at any time during the day. At a distance of 23.44229 degrees from the South Pole, the Sun will just touch the horizon at noon. (Again we ignore the effect of atmospheric refraction.)

In general, under these conditions, the whole northern hemisphere will be getting more daylight than darkness; and the whole southern hemisphere will be getting more darkness than daylight.

The situation as I've just described it is what exists at

the summer solstice, which takes place at June 21 on our calendar.

As the Earth revolves about the Sun, however, the direction of the axis relative to the stars does not change. Half a year after the summer solstice, when the Earth is at the other end of its orbit, the axis is slanted so that the North Pole is tipped directly away from the Sun. At that time, at December 21, the winter solstice, the situation is exactly as described for June 21 except that north and south have switched places.

At the winter solstice it is the South Frigid Zone which is in twenty-four-hour daylight and the southern hemisphere, generally, which gets more light than darkness, while the North Frigid Zone is in twenty-four-hour dark and the northern hemisphere, generally, gets more darkness than light.

From June 21 to December 21, as the Earth revolves, days grow shorter and nights longer in the northern hemisphere, while days grow longer and nights shorter in the southern hemisphere. From December 21 to June 21 the situation is reversed. For half a year, centered on June 21, the northern hemisphere gets more daylight and solar heat than does the southern hemisphere. For half a year, centered on December 21, the situation is reversed.

It is the tipping of the axis, then, that causes the seasons. The months centered about the summer solstice comprise spring and summer in the northern hemisphere, fall and winter in the southern hemisphere.† For the months centered about the winter solstice, it is fall and winter in the northern hemisphere, spring and summer in the southern hemisphere.

This unevenness evens out over the course of the year. The changes are just about symmetrical north and south, and in the long run, every spot on Earth's surface gets roughly equal amounts of darkness and light. (Actually, as a result of atmospheric refraction, every spot on Earth's surface gets a little bit more light than darkness and this unevenness becomes more marked the nearer we are to the poles.)

Light, however, is not equally effective everywhere. The farther from the Equator we are, the lower in the sky the

† To call June 21 the *summer* solstice is northern-hemisphere chauvinism.

Sun is, on the average, and the less heat is delivered per unit surface. On the whole, then, the average local temperature goes down as we move away from the Equator, north or south.

As it happens, the Earth is a very watery planet and its average temperature, as a planet, is not very far above the freezing point of water.

As we go farther and farther from the Equator, north or south, and as the average local temperature drops, it becomes more and more likely that the temperature will drop low enough to freeze water. Around each pole, therefore, there is ice. During the half year centering about the winter solstice, the ice tends to advance in the far north and retreat in the far south. During the half year centering about the summer solstice, the ice tends to retreat in the far north and advance in the far south.

The axial tilt therefore results in a pendulumlike swing of the ice, the swing being in opposite phases in the two hemispheres.

Yet the swing appears to be in equilibrium. Each advance is to roughly the same point in the winter, each retreat to the same point in the summer. The amount of freezing that takes place in the winter is just balanced by the amount of melting that takes place in the summer and on the whole the ice remains within bounds.

But will that always be so? What happens if, for some reason, just a little more freezing takes place in the winter than melting does in the summer? Each year a little more ice would then accumulate than had existed the year before and the world might, little by little and year by year, become far more icy than it is now.

Could that happen? Yes, it could. We know that it could happen in the future because it *has* happened in the past—and a number of times. There have been Ice Ages recurring with a kind of periodicity.

Why should there be Ice Ages? If things are in balance now, what could there be that would throw them off balance?

Could the Sun, for some reason, grow cooler at times? There's no evidence for that. Could the Sun have passed through regions in space in which dust was thicker and absorbed solar heat before it reached the Earth? There's no evidence for that, either.

So let's consider the Earth's orbit further and see if there is any unevenness to it at all.

If the Earth moved about the Sun in a perfect circle, with the Sun at the exact center of the circle, then the Earth would remain at a constant distance from the Sun at all times. Barring changes in the Sun itself or in space around the Sun, the Earth would, in that case, receive heat at a constant rate from the Sun.

This, however, is not so. As was first pointed out by the German astronomer Johannes Kepler, in 1609, the Earth moves about the Sun in an ellipse.

An ellipse can best be described, nonmathematically, as a kind of flattened circle (see Figure 2).

In a circle, every diameter (that is, any straight line passing through the center from one side of the circle to the other) is equal in length. In an ellipse, the diameters vary in length. The shortest diameter is from one flattened side to the other flattened side and that is the "minor axis." At right angles to the minor axis is the longest diameter, which is the "major axis." Where the major and minor axes cross is the center of the ellipse.

On the major axis there are two points called the "foci" (singular, "focus"), one on one side of the center and the other an equal distance on the other side of the center.

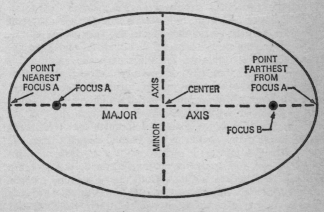

POINT NEAREST FOCUS A    FOCUS A    AXIS    CENTER    POINT FARTHEST FROM FOCUS A

MAJOR    AXIS

MINOR

FOCUS B

Figure 2 — The Ellipse

One property of the ellipse is this: if a straight line is drawn from one focus to any point on the curve of the ellipse, and from that point another straight line is drawn back to the other focus, the total length of the two lines is always the same and is always equal to the length of the major axis.

If we concentrate on one of the foci (call it Focus A), then we find that its distance from the curve of the ellipse changes continually as we mark out that curve. The part of the ellipse which is closest to Focus A is the end of the major axis on the same side of the center. The part of the ellipse which is farthest from Focus A is the end of the major axis on the other side of the center.

The more flattened the ellipse, the farther the two foci are from the center and from each other.

If the ellipse is flattened only very slightly, then the two foci are closer to the center and to each other. The difference in distance from a focus to the near end of the major axis and from that same focus to the far end of the major axis is not, then, very great. If the ellipse is very flattened, the two foci are widely separated from the center and each other, and are very near the opposing ends of the ellipse. In that case, each focus is very close to the near end of the major axis and very far from the far end. The difference in distances from a focus to various parts of the ellipse is, in that case, enormous.

Another way of looking at it is this—

The more flattened an ellipse is, the farther apart the foci are and the closer they are to the ends of the ellipse. Therefore, the more flattened an ellipse is, the larger the distance between the foci as compared to the length of the major axis. The ratio of the distance between the foci to the length of the major axis is called the "eccentricity" (from Greek words meaning "out of center").

When an ellipse is infinitesimally flattened, the foci are an infinitesimal distance from the center and from each other so that the eccentricity is virtually equal to zero. If the ellipse is flattened into something that is only infinitesimally removed from a straight line, the foci are only an infinitesimal distance from the ends of the straight line and the eccentricity is virtually equal to one. For any actual ellipse, the eccentricity lies between 0 and 1, and the smaller the value the nearer the ellipse is to a circle.

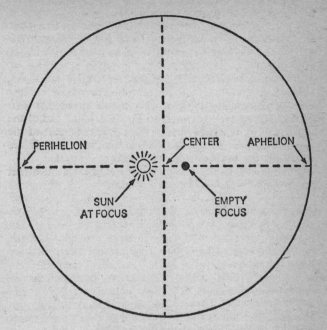

*Figure 3 — Perihelion and Aphelion*

What has all this to do with Earth's orbit about the Sun?

Well, not only is the orbit an ellipse, but the Sun is located not at the center, but at one of the foci. That means that if you imagine a line drawn along the major axis of the Earth's elliptical orbit, the Sun will be on that line, but closer to one end of the ellipse than to the other (see Figure 3).

When the Earth passes the end of the major axis which is on the same side of the center as the Sun focus is, the Earth is then at a minimum distance from the Sun. It is then at "perihelion" (from Greek words meaning "near the Sun"). Six months later it is at the other end of the major axis, and it is farthest from the Sun. It is then at "aphelion" (from Greek words meaning "away from the Sun").

Fortunately for life on Earth, the eccentricity of Earth's orbital ellipse is not very high. It is, in fact, only 0.01675, and if you drew an ellipse with that exact eccentricity, you could not tell, by eye, that it was not a circle.

Still, in an ellipse as enormous as that of Earth's orbit, even a small eccentricity is large in terms of kilometers. The major axis of Earth's orbit is 299,000,000 kilometers (185,800,000 miles) long, and the two foci are separated from each other by 5,002,000 kilometers (3,108,000 miles).

At perihelion, then, the Earth is 5,002,000 kilometers (3,108,000 miles) closer to the Sun than it is at aphelion. At perihelion the Earth is 147,000,000 kilometers (91,-350,000 miles) from the Sun, while at aphelion, the Earth is 152,000,000 kilometers (94,450,000 miles) from the Sun.

That's a difference of about 3.3 per cent, which isn't very much really. It means the apparent orb of the Sun is slightly larger at perihelion than at aphelion, but not enough to be noticed by nonastronomers. It means that the Earth moves faster in the perihelion half of the orbit than in the aphelion half so that the seasons are not of exactly equal lengths, but again who would notice?

Finally, though, it means that at perihelion we get more heat from the Sun than at aphelion. The heat we get varies inversely as the square of the distance, so that it turns out Earth gets almost 7 per cent more heat at perihelion than at aphelion.

Let's look at it this way. Halfway between perihelion and aphelion (at one end of the minor axis of the ellipse) the Earth is at about its average distance from the Sun and is receiving about an average amount of heat.

If it is then moving toward the perihelion, through the perihelion and back to the other end of the minor axis, then during that half of Earth's orbit. Earth is receiving more than average heat from the Sun, with a maximum of a little over 3 per cent above average at perihelion.

As the Earth then moves through the aphelion and back to the starting point, it then receives less than average heat in that half of the orbit, with a minimum of a little over 3 per cent below average at aphelion.

Does this matter?

It wouldn't, if the Earth's axis were perfectly upright since both northern and southern hemispheres would then share equally in the shifting heat-receipt in the course of the year. But the axis is tilted; how does that affect things?

Earth reaches its perihelion on January 2‡ and its aphelion on July 2. It so happens that January 2 is less than two weeks after the winter solstice, while July 2 is less than two weeks after the summer solstice.

This means that at the time the Earth is at or near perihelion, and getting more heat than usual, the northern hemisphere is deep in winter, while the southern hemisphere is deep in summer. The extra heat means that the northern winter is milder than it would be if the Earth's orbit were circular, while the southern summer is hotter.

At the time the Earth is at or near aphelion and getting less heat than usual, the northern hemisphere is deep in summer, while the southern hemisphere is deep in winter. The heat deficiency means that the northern summer is cooler than it would be if the Earth's orbit were circular and the southern winter is colder.

We see, then, that the combination of Earth's orbital ellipticity and its axial tilt produces an asymmetry. The northern hemisphere has a less extreme swing between summer and winter than the southern hemisphere does. The difference is not much, but it is there.

This might be taken as meaning that the advance and retreat of ice in the far south is more extreme than in the far north. The colder winters of the far south mean a further advance of the ice than in the far north. The hotter summers of the far south mean a further retreat of the ice than in the far north. You might feel that this means it is the southern hemisphere that is, at the moment, in greater peril of an Ice Age than the northern hemisphere is.

If you think that, you're wrong. It is actually the less extreme swing which encourages an Ice Age, for changes in ice accumulation are more sensitive to changes in summer temperature than in winter temperature.

Thus, a slightly lower average winter temperature does not necessarily mean more snow, nor does a slightly higher average winter temperature necessarily mean less snow. The reverse is more likely true. A slightly higher winter temperature (but one that is still below the freezing point)

‡ That's my birthday, but I don't suppose there's any connection.

means more water vapor in the air and therefore *more* snow.

On the other hand, a slightly lower summer temperature means less melting, and there's no alternative to that.

The northern hemisphere, with its slightly milder winters and cooler summers, therefore tends to have more snow and less melting than the southern hemisphere does, so that if there is any danger of an Ice Age, it is in the northern hemisphere rather than in the southern.

But the northern hemisphere has had Ice Ages in the past. If the summer-winter temperature swing tends to favor Ice Ages right now, why did the ones in the past stop? Are there mitigating circumstances now?

Well, in the course of the half of Earth orbit that is closer to the Sun, the Earth, feeling a stronger gravitational pull, moves a bit more quickly than it does in the other half. That means it takes Earth about 186.5 days to go from one end of the minor axis of its orbital ellipse through aphelion to the other end of the minor axis. It takes Earth only about 178.8 days to go from that other end of the minor axis through the perihelion back to the first end.

In the northern hemisphere, fall and winter take place during the perihelion half of the orbit and are only 178.8 days long, taken together. It is the spring and summer which are in the aphelion half of the orbit and which are 186.5 days long altogether.

The northern winter, then, which has the potentiality of producing more snow because it is slightly milder than the southern winter, is cut a little short and therefore does not produce as much snow as it might if the seasons were all equal in length. The northern summer, which is short on ice-melting potentiality because it is slightly cooler, gets in a little more work at it than it otherwise would if the seasons were equal in length.

The result is that the situation, north and south, is not as asymmetric as one might suppose. Or, at least, one asymmetry tends to cancel out the other asymmetry.

The canceling is not complete. The slightly cooler summers of the northern hemisphere still encourage the Ice Age despite their somewhat greater length.

Which leaves us still with the question of why there are Ice Ages at some times and not at others. If the combination of axial tilt and elliptical orbits is enough to produce an Ice Age in the northern hemisphere, why doesn't it produce one? If it is not enough to produce an Ice Age, why have Ice Ages occurred in the past?

Ah, but we're not through yet with the peculiarities of Earth's orbit. The orbit does not repeat itself exactly through all eternity. For that matter, neither is Earth's axial tilt fixed for all eternity.

Both orbit and tilt would be fixed if Earth and Sun were alone in the Universe, but they are not alone. The Moon is present also, as are the planets and even the distant stars. Each of these other bodies has a gravitational field and each of those gravitational fields has the potentiality of influencing the Earth's motion.

All these other bodies are much smaller than the Sun, or much more distant than the Sun, or both, so that none can, in any way, compete with the Sun's overwhelming gravitational effect on Earth. Despite all the pulls in the Universe, therefore, Earth continues its stately motion about the Sun, almost unaffected by the other objects in existence.

*Almost unaffected.* Not completely unaffected.

The extraneous pulls to which the Earth is subjected produce minor changes in the Earth's orbit (perturbations), all of which are very small indeed over ordinary time spans so that they don't affect ordinary human affairs at all in the space of a lifetime or bother anyone but astronomers.

Even very small perturbations can, in the long run, produce effects out of all proportion to their size, however, and it is in tiny perturbations that the secret of the Ice Ages is now thought to rest.

And it is those perturbations we'll take up in Chapter 8.

## Eight
# OBLIQUE THE CENTRIC GLOBE

I suppose you are sophisticated enough to know that these articles are not written the day before the book containing them hits the stores. They are written months before. Although you may be reading this chapter deep in the summer, for instance, and are experiencing, perhaps, a heat wave, it was written deep in winter.

In fact, shortly after I wrote Chapter 7, dealing with new material recently broached concerning the Ice Ages, the eastern two thirds of the United States went into what its shivering inhabitants consider Ice Age enough.

Although I am insensitive to cold (within limits), even I had to agree enough was enough on the morning of January 17, 1977, when I was waiting at a suburban Philadelphia railroad station for a train to come and take me home to New York. I had arrived at 6:05 A.M. for a train that was due at 6:40 A.M. (I'm an early bird) and that train arrived at 7:30 A.M. The temperature was below zero Fahrenheit, and although I waited inside a reasonably warm waiting-room with a couple of dozen other people, the *idea* of cold without made all of us miserable.

At least it puts me in the mood to continue the discussion.

In Chapter 7 I pointed out that the ellipticity of the Earth's orbit and the tilting of the Earth's axis combined to produce mild winters and cool summers in the northern hemisphere and cold winters and hot summers in the southern hemisphere. I also explained that it was the mild-winter-cool-summer situation that tended to produce Ice Ages and the question was: why isn't the northern hemisphere experiencing an Ice Age now, then?

Well, let's see—

The Earth turns on its axis, and any turning object, as a result of its inertia (the tendency of any point on its surface to move in a straight line rather than in a circle) experiences a centrifugal effect that tends to move every part of it away from the center of rotation.

Because the Earth is a ball that turns all in one piece, different parts of it turn at different rates. At the North and South Poles the surface is located right on the axis of rotation and there is no rotational motion at all. The farther one goes from the poles, the more rapid the motion of the surface (and, on the whole, of the material under the surface, too) until we reach the Equator where the motion is most rapid—a point on the Earth's surface at the Equator has a rotational speed of 27.83 kilometers per minute (17.29 miles per minute).

The centrifugal effect increases, then, from zero at either pole to a maximum at the Equator. The Earth bulges away from the axis of rotation by an amount that increases steadily as one moves from either pole and reaches a maximum at the Equator. The bulge is therefore called an "equatorial bulge" and it is 22 kilometers (13 miles) high at the Equator.

If the Earth were exactly spherical, the gravitational pull of other bodies upon it would act as though it were exerted entirely on Earth's center. Because of the equatorial bulge, the Earth is not exactly spherical so that there is an additional pull on the gravitational centers of the bulge (one on each side of the Earth) in addition to the pull on the center.

If the Moon revolved about the Earth exactly in the equatorial plane, this would not matter. The gravitational center of the Earth and of the equatorial bulge both on the side toward the Moon and on the side away from the Moon would all be in a straight line, and the bulge would then not introduce any complications.

The Moon, however, revolves in a plane markedly tilted to Earth's equatorial plane. That means that the Moon pulls on the three centers of gravity in slightly different directions and at slightly different distances.

The effect is to make the Earth "precess."* That is, without the extent of the axial tilt changing, the North and South Poles each describe a circle relative to the imaginary line that is perpendicular to Earth's orbital plane about the Sun.

(We can see this happen when a top is spinning. If it is

---

* The Sun's pull also plays a role, but a lesser one.

tilted as it spins, the Earth's pull on it causes it to wobble, so that its tilt turns about the point it is spinning upon. Of course, Earth isn't spinning on a point, so both ends of its axis wobble about a fixed point at the center of the axis.)

If the Earth's axis is extended, in imagination, to the sky, the North Pole and South Pole impinge upon the sky at the North Celestial Pole and the South Celestial Pole. We can tell the location of these celestial poles because the rest of the sky turns about them.

If we watch from year to year and from decade to decade we find that the position of the North and South Celestial Poles change slowly, as a result of the precession of Earth's axis. In fact, each celestial pole marks out a circle about 47° in diameter, completing one turn about the circle in 25,780 years.

And what happens to Earth's orbit as a result of precession?

At the present moment, the north polar end of the axis is tipped most toward the Sun on June 21, at which time Earth is nearly as far from the Sun as it can get, which is why the northern hemisphere summers are cooler than they would otherwise be and the southern hemisphere winters are colder. The north polar end of the axis is tipped most away from the Sun on December 21 at which time Earth is nearly as close to the Sun as it can get, which is why the northern hemisphere winters are milder and the southern hemisphere summers are hotter than they would otherwise be.

But (assuming all else remains fixed) in 12,890 years precession will have turned the axis so that it is tilting in the opposite direction. On June 21, when Earth is far from the Sun, the north polar end of the axis will be tipped away from the Sun, and on December 21, when Earth is near the Sun, the north polar end of the axis will be tipped toward it.

The situation will be precisely the opposite to what it is now. It will be the northern hemisphere that will be getting cold winters and hot summers and the southern hemisphere that will be getting mild winters and cool summers. It will be the southern hemisphere that will be threatened with an Ice Age, and not the northern.

We cannot, of course, deal with precession alone, because the perihelion doesn't stay in the same place. If the Earth and Sun were alone in the Universe, the Earth's orbit would be a closed ellipse, the Earth would repeat its path about the Sun exactly for an indefinite period of time, and the perihelion would stay put.

But the Earth and Sun are not alone and the result of extraneous gravitational pulls on the Earth brings about complications.

If the Earth is imagined to start its orbit at perihelion, it does not reach the same point in space (relative to the Sun) when it returns to perihelion. If the Earth, as it moved around the Sun, left a mark, you would see it describe not a closed ellipse but a kind of complicated rosette, each turn cutting space in a slightly different line.

The net effect of all this is that the perihelion point slowly moves about the Sun, so that the Earth reaches it at a slightly different place and time each year. The perihelion makes a complete circle about the Sun in about 21,310 years. Every fifty-eight years, the day of perihelion shifts by one day on our calendar.

Therefore the question of which hemisphere has mild winters and cool summers and which has cold winters and hot summers depends on the combined effect of precession and perihelion movement.

In 1920 a Yugoslavian physicist, Milutin Milankovich, suggested that there was a great weather cycle as a result of small periodic changes involving the Earth's orbit and its axial tilt. He spoke of a "Great Winter," during which the Ice Ages took place, and a "Great Summer," which represented the interglacial periods. In between would be a "Great Spring" and a "Great Fall," of course.

If we considered only the precession-plus-perihelion motion, we might suppose that when the northern end of the axis was most tipped toward the Sun at perihelion, the northern hemisphere would have the hot-summer-cold-winter combination at its most extreme. That would be the Great Summer solstice for the northern hemisphere, the June 21 of the Great Season. At such a time, of course, the southern hemisphere would be experiencing the cool-summer-mild-winter combination at its most extreme and that would be the Great Winter solstice for it, the December 21 of the Great Season.

When the axis tilts in the opposite direction at perihelion, it would be the Great Winter solstice for the northern hemisphere and the Great Summer solstice for the southern hemisphere.

At the present moment we are very close to the Great Winter solstice for the northern hemisphere. Why, then, don't we have an Ice Age, here?

For one thing, perhaps because there's a natural lag.

December 21 in the ordinary year may be the winter solstice and may be the time of the shortest day and the longest night of the year, but it is not likely to be the coldest day of the year. It is, in fact, only the beginning of winter.

After the winter solstice, the days are growing longer and the nights shorter, but for quite a while, the days remain shorter than the nights so that there is a continuing deficit of heat, with more being lost at night than is gained during the day from the Sun. As a result, the temperature continues to drop, on the average, through January and into early February, which is the depth of winter. (In the same way, the temperature continues to rise, on the average, past the summer solstice on June 21, through July and into early August.)

In the same way, the Great Seasons may lag as the effect accumulates past the solstice. If someone in December said, "Where's the snow?" the answer would be "Wait!" —and so it might be now.

If it were only a matter of precession-plus-perihelion motion, Ice Ages would come in the hemispheres alternately. The depth of an Ice Age in the northern hemisphere would be at the height of an interglacial period in the southern hemisphere and vice versa. There is evidence, however, that Ice Ages take place in both hemispheres simultaneously.

There may be other effects, then, that do their work on both hemispheres, and it may be these other effects that predominate over precession-plus-perihelion motion.

For instance, one effect of various gravitational pulls on Earth is to cause the axial tilt to wobble, not just in precession fashion but in actual amount.

At the present moment, the axial tilt is 23.44229 degrees to the Earth's orbital plane, but this is not immutable. It is decreasing. Back in 1900 it was 23.45229 degrees and in 2000 it will be 23.43928 degrees.

If this decrease continued steadily over the centuries, then in 137,000 years the axis would be bolt upright and the seasons would vanish. Of course, that won't happen.

The present decrease in the amount of axial tilt is part of a cycle, back and forth. It will reach a minimum of not very much less than the present value—about 22 degrees—and then increase to a maximum of not very much more than the present value—about 24½ degrees—and then repeat this over and over again indefinitely. The length of the cycle is 41,000 years.

How does this affect the Earth's weather? —Not the way most people seem to think.

We all know that it is because of the axial tilt that we have winter and summer. If there were no axial tilt at all, there would be days and nights equal in length over all the world. The situation would be permanently what it is now at the equinoxes.

It seems natural, then, to have the idea that if only the Earth's axis were not tilted, there would be an eternal spring everywhere on Earth.

This idea finds expression in *Paradise Lost* by John Milton (who was great on poetry but weak on astronomy). Milton felt that before the Fall, when man still lived in Eden, there was no axial tilt and there was a world-wide and eternal spring. It was only after the Fall that the tilt was imposed.

Milton, who wanted to cling to the Ptolemaic theory but reluctantly recognized the fact that astronomers were, by the time he was writing, virtually all Copernicans, wasn't sure whether to say the tilt came about by tipping the Earth or tipping the Sun—so he waffled. In Book X of his epic, he writes:

> Some say he [God] bid his Angels turn askance
> The Poles of Earth twice ten degrees and more
> From the Sun's Axle; they with labour push'd
> Oblique the Centric Globe: Some say the Sun
> Was bid turn Reins from th' Equinoctial Road
> Like distant breadth . . .

Milton was wrong, however, in thinking of the tilt (imposed either Copernically or Ptolemaically) as a punishment.

Suppose the axis were tilted less than it now is. In that case, the unevenness in length of day and night in the regions about the solstices would be less. The summer wouldn't be so hot or the winters so cold. There would be a mild-winter-cool-summer for *both* hemispheres. The less the axis was tilted, the milder the winter and the cooler the summer for *both* hemispheres.

However, as I explained in Chapter 7, a mild winter tends to produce more snow and a cool summer to melt less snow. A smaller tilt to the axis encourages an Ice Age in both hemispheres, therefore, and if the axis were not tilted at all, the Ice Age would be permanent, north and south.

So tilting the axis was a reward, in that it unfroze the world.

In fact, one could argue this way. As long as Adam and Eve were in the Garden, which we might picture as in a tropical clime, a seasonless year was beneficial. After the Fall, when human beings were going to multiply and spread out over the world, the Temperate Zones would have to be made habitable for them and hence the tilt was imposed. Had Milton been able to advance this explanation, he could have illustrated God's loving-kindness rather than His vengeance—which means he would probably not have talked of the tilt at all, for pious people, in my experience, are more interested in vengeance.

Anyway, the point is that the axial tilt is currently in the decreasing stage and that favors the coming of the Ice Age for both hemispheres.

We're still not through.

At the time of writing the eccentricity of Earth's orbital ellipse is 0.01675 and the difference in the distance from the Sun at perihelion and at aphelion is 5,002,000 kilometers (3,108,000 miles), or about 3.3 per cent of the average distance.

That eccentricity wobbles, too, in a cycle of 92,400 years. The eccentricity can decrease to 0.0033, or only ⅕ its present amount, then increase to a maximum of 0.0211, or 1¼ its present amount.

At eccentricity maximum the Sun is 6,310,000 kilometers (3,920,000 miles) closer at perihelion than at aphelion. At eccentricity minimum, the Sun is 990,000 kilometers (610,000 miles) closer at perihelion than at aphelion.

The less the eccentricity and the more nearly circular the orbit, the smaller the difference in the amount of heat the Earth gets from the Sun at different times of the year.

This decreases the cold-winter-hot-summer chances and encourages the mild-winter-cool-summer situation.

In other words, a period of declining eccentricity is a period that tends to favor the approach of Ice Ages, and, as it happens, Earth's orbital eccentricity is declining right now. The eccentricity is decreasing now at a rate of 0.0004 per century. In other words, each year (right now) Earth, at perihelion, is 1.2 kilometers (0.75 miles) farther from the Sun than at the previous perihelion.

All these orbital and axial changes are small and it is odd that they can produce the enormous changes in ice cover that they do. The reason for that is that the advance and retreat of the ice involves a vicious circle (if you disapprove of the change) or a beneficial one (if you approve of it).

Suppose that Earth's orbital and axial wobbling produces a seasonal change that encourages a slight expansion of the ice cover. It so happens that ice reflects light much more efficiently than does liquid water or bare soil. The fact that there is more ice, therefore, means that more sunlight is reflected by the Earth as a whole and less is absorbed than before. That drops the average temperature of Earth and encourages still more ice formation, which lowers the temperature further, encourages still more, and so on.

In the end, a small expansion of the ice cover can trigger enormous ice sheets and a semiplanetary freeze.

It works the other way in the depth of the Ice Age. If orbital and axial variations produce a small retreat at the edges of the ice sheet, less sunlight is reflected, there is a slight rise in Earth's average temperature, which encourages further ice retreat, further temperature rise, and so on.

In the end, a small retreat of the ice sheet can trigger the melting of the whole and restore a mild-temperatured Earth.

It would seem, then, that if one developed a way of measuring the temperature of the Earth with all its tiny variations, one might find a complicated, but regular pattern, which can be shown to be made up of the various cyclical wobblings of orbit and axis. If one did, that would be strong evidence that those wobblings had important effects on Earth's temperature, effects that could only be produced by way of Ice Ages.

The problem was tackled by J. D. Hays (Columbia University), John Imbrie (Brown University), and N. J. Shackleton (Cambridge University) and their results were published in December 1976.

They worked on long cores of sediment dredged up from two different places in the Indian Ocean. The places were far from any land areas so there would be no material washed from the land to obscure the record. The places were also relatively shallow so that there would be no material washed down from surrounding, less deep, areas.

The sediment, it could be supposed, would be undisturbed material laid down, on the spot, for century after century, and the length of core brought up stretched backward, it seemed, over a period of 450,000 years. The hope was that there would be changes as one went along the cores that would be as distinctive and as interpretable as tree rings.

But that meant there would have to be something in the sediments that would do the job of tree rings. By the spacing of the tree rings, one could identify wet summers and dry summers. What was there in the sediment that could identify warm periods and cool periods? What would serve as a thermometer?

Actually, there were two thermometers—two very different, and independent ones—so that if the two agreed, that would be significant.

The first involved the tiny Radiolaria that lived in the ocean through all the half-million years being investigated. These are one-celled protozoa with tiny, elaborate skeletons which, after the deaths of the creatures, drift down to the sea-bottom as a kind of siliceous ooze.

There are numerous species of Radiolaria, some of which flourish under warmer conditions than others. They are easily distinguished from each other by the nature of their skeletons, and one can therefore poke along the cores of sediments, millimeter by millimeter, studying the nature of the radiolarian skeletons and estimating from that whether, at some given time, the ocean water was warm or cool. One could, in this way, set up an actual curve of ocean temperature with time.

The second thermometer involved not living things but atoms. Oxygen consists predominantly of oxygen-16 atoms. About one oxygen atom out of five hundred, however, is oxygen-18. (There are also a very few oxygen-17 atoms about, but their presence doesn't affect the following argument.)

The oxygen-18 atoms are 12.5 per cent more massive than the oxygen-16 atoms. A water molecule containing oxygen-18 has a molecular weight of 20, compared to 18 for a water molecule containing oxygen-16. That is an 11.1 per cent weight difference.

When solar heat evaporates water from the ocean, water molecules containing oxygen-16, being lighter, vaporize slightly more readily than do those containing oxygen-18 atoms. At any given time, the water vapor in the atmosphere and the rain it condenses to are richer in oxygen-16 and poorer in oxygen-18 than the water of the ocean is.

The disparity doesn't ordinarily build up. The water vapor condenses to rain and falls into the ocean again, or falls on land and, in not too long a time, flows into the ocean.

In the course of an Ice Age, however, a great deal of the water vapor ends up as snow, which lands on the growing icecaps and remains there, not returning to the oceans for tens of thousands of years.

The ice sheets represent a huge reservoir containing water molecules which are rich in oxygen-16 and poor in oxygen-18. The more voluminous Earth's mass of ice, the greater the quantity of oxygen-16 preferentially withdrawn and the higher the percentage of oxygen-18 in the water of the still-liquid ocean and in any molecules that incorporate the oxygen of that water.

One can therefore go along the sediment core, millimeter by millimeter, determining the oxygen-18: oxygen-16 ratio. The higher the ratio, the more advanced the Ice Age and the lower Earth's temperature.

Both thermometers yielded just about identical results in both cores. What's more, the temperature curves obtained could be shown to be built of simple cycles closely resembling those that would be expected of the known orbital and axial variations of Earth.

There seems to be good reason to think, then, that it is the orbital and axial variations that are indeed the cause of the Ice Ages and that the curve we obtain may be used to predict the future in this respect.

We appear right now to have passed one of the pronounced peaks in the curve that recurs at 100,000-year intervals and which represents mild interglacial conditions and to be heading downward toward a new Ice Age.

That doesn't mean next year, of course (though those of us who have just lived through the frigid January of 1977 might be excused for having our pessimistic doubts), or even next millennium. Nevertheless, however far in the future the next Ice Age may be set, there is still cause for concern right now.

Long before the cooling conditions are severe enough to bring about the southward grinding of the glaciers, they will be enough to shorten the growing season a bit and to increase the incidence of killing frosts early and late in the season at the northern and altitudinal edges of a particular crop region. Good harvests will be bound to become less frequent, and this, combined with increasing population (if it continues to increase), will make mass starvation the more certain.

Is there anything we can do about it? Perhaps, yes. The scientists working with the cores specifically state that the temperature curve does not allow for "anthropogenic" effects.†

Humanity is doing things which were never before done in the course of the 450,000-year period over which the curve has been worked out. Humanity has been burning fossil fuels at an increasingly rapid rate and has been pouring carbon dioxide into the atmosphere in unprecedented quantities. This will not serve to change the natural percentage of carbon dioxide in the air very much, but it could be enough to increase the greenhouse effect (see "No More Ice Ages?" in *Fact and Fancy*, Doubleday, 1962) sufficiently to abort an Ice Age.

Then, when the fossil fuels are used up and humanity turns to other sources of energy such as nuclear fusion and space solar-power stations, the heat developed in this way and added to the supply we naturally get from the Sun may continue to abort Ice Ages and, indeed, subject us to the risk of overheating, with consequent melting of the ice sheets that yet remain in Greenland and Antarctica and catastrophic flooding of the continental lowlands.

Yet one thing remains to be explained—

If, indeed, orbital and axial changes are the cause of the

---

† Don't let the word throw you. It's just Greek for "man-made," and scientists use it only to irritate typesetters.

Ice Ages, then such Ice Ages should have occurred periodically over all of Earth's history. Instead, they seem to have been taking place only during the last million years. Before then there were about 250,000,000 years without serious Ice Ages.

Secondly, the temperature curve would seem to show that the two polar regions are equally affected, yet it is the northern hemisphere that has suffered from Ice Ages almost exclusively.

There is some source of asymmetry both in time and in space, and it cannot be in the orbital and axial changes, so it has to be somewhere else—and that will be the topic for Chapter 9.

## Nine
# THE OPPOSITE POLES

As it happens, I am very accessible. It is not at all difficult to dig up my address or my phone number. I make no particular secret of them. I have no desire to withdraw or to hide.

This creates alarm and despondency in the hearts of those near and dear to me, however, for they have visions of my being bothered to death by all kinds of well-meaning (or eccentric) individuals.

In response, I explain that I have faith in my Gentle Readers. They are, by and large, judging by those I have seen and heard from, an intelligent and considerate lot, who do not take undue advantage of me. My letter box is usually full; my phone rings often; but both letters and phone calls are almost invariably reasonable ones, not overlong, and make not too many demands.

But then, every once in a while—

Not very long ago, the phone rang at 3:30 A.M. What's more, it was not the particular phone which is in my bedroom that rang, but another one several rooms away. At that time of night and on that phone, I expected disaster. I assumed it was one of my children and I further assumed it was a dire emergency.

Fortunately, I sleep lightly and I wake quickly. Patter, patter, patter, went my bare feet on the floor, and I was at the phone.

"Hello," I said, in breathless apprehension.

"Dr. Asimov?" came the eager voice of a stranger.

"Yes. Who is this?"

Still eager: "I want to speak to you, Dr. Asimov, and ask—"

"Wait a while. Do you know it is 3:30 A.M.?"

There was a slight pause as though the stranger stopped to wonder a bit at my reason for introducing so irrelevant a fact. "Yes, of course," he said.

"Why do you call me at 3:30 A.M.?" I asked.

He said, "I'm a night owl," as though surprised I didn't know.

And I answered in the same tone of voice, "And I'm not," and hung up on him. It was rude of me, but I really felt justified.

That some people are night owls and some are not is a truism, and I was rather chagrined to think that somewhere among my readers is a young man so idiotic as to fail to realize that this pair of opposites exists and to assume that his own personal peculiarities are standard for the whole world.

But a writer can find something useful in anything. Musing on opposites before I drifted off to sleep again, I found my writing strategy for this chapter.

In the previous two chapters I discussed the small, periodic, astronomic changes that might possibly cause the periodic Ice Ages on Earth. However, the changes in Earth's motions have been going along for uncounted millions of years, presumably, while the recent periodic Ice Ages have been a matter only of the last million years on Earth. Before that, there was something like 250,000,000 years without Ice Ages, as nearly as we can tell.

Somewhere there is an asymmetry, and if we look, we can find it on Earth's surface. There we find (aha!) opposites. These opposites are, of course, land and sea—solid and liquid—stay-put and flowing. And those opposites are *not* symmetrically distributed. Given that, let us see how this might affect the Ice Ages.

We can begin by supposing Earth's surface to be symmetrical with respect to these opposites.

Suppose, for instance, that the Earth's land surface was restricted to the Tropical Zone. We would have a belt of land (possibly broken by narrow arms of the sea) around Earth's middle and a broad and unbroken ocean taking up the Temperate and Frigid Zones on both sides of the land, north and south. Earth would then have two polar oceans, and the land-sea distribution would be symmetrical.

At either pole, freezing would take place. Sea water freezes at a temperature of $-2°$ C. ($29°$ F.) and the conditions at the poles will produce temperatures lower than this in the wintertime.

An icy layer will cover those portions of the ocean,

then, and it may cover an area as great as 13,000,000 square kilometers (5,000,000 square miles) at its maximum extent in the depth of winter. This is nearly 1.5 times the area of the United States. In the summer, much of the ice layer will melt and no more than perhaps 10,000,000 square kilometers (3,900,000 square miles) will remain covered.

The two poles will alternate in this matter. When the North Pole has its maximum cover of solid ice, the South Pole will have its minimum cover, and vice versa.

At any one moment, an Earth with polar oceans will have some 23,000,000 square kilometers (8,900,000 square miles) covered with ice. This amounts to about 4.5 per cent of the Earth's surface.

This sea ice, however, would not be very thick. Ice is a good insulator and once some of it has formed, the water underneath the layer loses heat only slowly through the ice overhead and therefore freezes only very slowly. The thicker the ice layer gets, the more slowly you can expect the water underneath to freeze.

The freezing process is slowed further because the ocean is a fluid and divided into currents that tend to equalize its temperature. Heat enters the polar ocean from the warmer tropics and this, too, limits the freezing under the ice layer.

To be sure, snow falls on top of the ice and that will add to its thickness, but that will also force the layer deeper into the water where some will melt.

After a winter of ordinary length and intensity the sea ice may end with an average thickness of no more than about 1.5 meters (5 feet) at its maximum in the dead of winter. The total cubic volume of ice on an Earth with two polar oceans would thus be about 34,500 cubic kilometers (8,300 cubic miles). This is about 1/35,000 of Earth's supply of water—an inconsiderable amount.

What happens in such a case if the small astronomic changes I discussed in the earlier articles produce a change in Earth's overall weather?

If the summer temperature cools somewhat, a little less ice melts, so a little more ice remains. This could trigger further change, since more ice serves to reflect more sunlight into space, thus cooling the summers a little further and encouraging still more ice formation.

This, however, would not go too far. The ocean circulation sees to it that there is a heat leak poleward from the tropics. The farther equatorward the ice creeps, the more effective is the leak, so that a new equilibrium is established that is not radically different from what existed before.

Consequently, the slow alternation of Ice Ages and milder periods between would see the polar ice expand somewhat and contract somewhat, over and over again. We might suppose that at the depth of an Ice Age, the thin layer of sea ice would extend an additional couple of million kilometers beyond the interglacial limits, and the effects of that would not be enormous.

We might even argue that the effects would be beneficial to life generally. Sea life depends, in part, on the amount of oxygen dissolved in sea water, and that amount increases as the temperature of the water decreases. The near-freezing cold water of the polar regions contains some 60 per cent more oxygen than the lukewarm water of the tropics. (It is for this reason that sea life is particularly rich in the polar oceans and why it is in coldwater currents that the world's great fisheries exist.)

On a planet with polar oceans, sea life would expand, prosper, and flourish during an Ice Age, and land life on the tropical continents might indirectly benefit, too.

One other point— Sea ice, present or absent, expanding or diminishing, would not affect the sea level. When water freezes to the slightly less dense form of ice, it expands. The frozen water, however, floats on the ice, with only a portion of itself submerged. The submerged portion is exactly equal in volume to that of the water which has frozen.

This means that land life on the tropical continents would be unaffected by the fact that far to the south and north, the sea ice was expanding and contracting—at least as far as the sea itself was concerned. The waters would neither slink down the beach with the years nor creep up it.

Now let us move to another kind of symmetrical condition. Let us leave most of Earth's continents in the tropics as before, but let's move some of the land to the poles. We will imagine Earth with two polar continents, each surrounded by a broad unbroken ocean.

To be specific let us suppose that each pole is occupied by a more-or-less circular continent, with the pole more or less at the center, and that the area of this continent is 13,100,000 square kilometers (5,000,000 square miles).

What would be the situation now?

We can suppose, to begin with, that the continents are bare; their surfaces simply exposed rock. In the polar winter the temperature of the bare surface would drop. It would drop faster than the temperature of water would in similar circumstances because the "specific heat" of rock is lower than that of water. An amount of heat loss that would reduce the temperature of a particular volume of water by 1 degree, would reduce the temperature of that same volume of rock 5 degrees.

Moreover, the temperature of sea water will sink only to its freezing point, and it will then freeze. The water beneath the ice will stay at that freezing point and will serve as a heat source that will prevent the ice above from cooling as much as it otherwise might.

The dry land of the polar continents, however, will cool as far down as the heat loss makes necessary and, since land does not flow, there are no land currents to bring heat in from outside. Therefore, the temperature of the land might, in the far interior and at the depth of winter, drop to below $-100°$ C. ($-148°$ F.).

In the polar summer, when the Sun may be low in the sky but shines for long periods of time (up to six months at a time at the poles themselves), the exposed land would rise in temperature to perhaps even mild levels—but the land doesn't remain exposed.

The surrounding ocean is the source of water vapor in the air and this can condense and precipitate (under polar temperature conditions) as snow. During the polar winter, snow will fall on the polar continent. Not very much will fall since the air will be too cold to hold much vapor, but some will, and the continent will receive a snow cover.

This means that in the polar summer, the Sun's heat will be devoted not to raising the temperature, but to melting the ice. It takes as much heat to melt a particular weight of ice as it would take to raise the temperature of that weight of water by 80 Celsius degrees (144 Fahrenheit degrees.)

This means that the polar continent remains cold during the summer and, in fact, that the Sun, hovering low in the sky, will not manage to melt all the snow that fell in the previous winter.

The next winter therefore sees a thicker snow cover, as new snow is added to the leftover of the previous winter; and the winter after that sees a still thicker one. There will, in the end, be a far thicker ice layer over land than, under the same conditions, over the sea.

Eventually, the layer of ice would be about 2 kilometers (1.25 miles) thick on the average, with a maximum thickness in the interior of about 4.3 kilometers (2.7 miles) perhaps. It would spread over an area of nearly 15,000,000 square kilometers (5,800,000 square miles), an area that would include the entire continent and some of the shallow inlets and bays along its shores.

The total quantity of ice would amount to 30,000,000 cubic kilometers (7,200,000 cubic miles) on each polar continent, or 60,000,000 cubic kilometers (14,400,000 cubic miles) altogether, and very little of it would melt in the summer.

The ice forming on polar continents would be nearly 1,750 times as great in amount as that forming on polar oceans. The polar continental ice would make up about 4.8 per cent of all the water on Earth.

Would the ice pile indefinitely on the polar continents until the entire ocean was on them in a precarious heap more than 40 kilometers (25 miles) high?

No. Ice is plastic under pressure, and when a few kilometers of it has piled up it tends to spread out like very stiff wax. Chunks of it (icebergs) split off the edge of the ice sheet and drift off into the surrounding ocean, amid what sea ice has formed there. This loss eventually balances the gain through snowfall so that the volume and thickness of the ice sheet reaches an equilibrium.

The icebergs are frozen fresh water, while sea ice traps brine between the freezing crystals and is therefore quite salty; fresh ice also melts less easily than salty ice does. Then, too, the icebergs are far thicker than the sea ice is and therefore take a longer time to melt. (Nevertheless, icebergs wander equatorward and melt eventually.)

In short, the ice surrounding a polar continent forms, together with the ice of the continent itself, a far greater reservoir of cold (or a far more effective heat-sink, if you prefer to look at it from the other direction) than a polar ocean does.

Hence a planet with a pair of polar continents would end up far more icy and far colder than the same planet with polar oceans, even though all the astronomical conditions were the same in either case.

What would Ice Ages be like on a planet with two polar continents and the remaining land in the tropics? Again, the change would not be much. The polar continents themselves could not very well be buried still deeper even if the summers grew somewhat cooler, but accumulating ice would accelerate the flattening out and cause the formation of icebergs at a greater rate. Therefore the sea surrounding the polar ocean would tend to get icier, again to a limited extent because of water circulation from the still warm tropics.

One thing, though. The ice piled up on a polar continent is removed from the sea, so that if affects the sea level. If, for any reason, the ice on the two polar continents were to melt entirely, the water would run off the continents into the sea and the sea level would be raised about 125 meters (420 feet). This would be a serious problem to land life.

However, such total melting would not be likely. The shifts in planetary weather between the Ice Age and the mild conditions between, on a planet in Earth's present astronomical situation, would not be great enough to affect seriously the polar continent's icecap one way or the other.

Thus, the Earth would not be affected much by Ice Ages if it had either two polar oceans or two polar continents, provided any other land were in the tropics.

You could have a compromise. There could be a polar continent at one pole only and a polar ocean at the other, with the remaining land in the tropics. You could, in this way, have two opposite poles, opposite not only in terms of geographic location, but in physical character.

In that case, the Ice Ages would still not have a great effect on Earth, but the polar asymmetry would produce a vast difference between the northern and southern hemispheres, as one pole and not the other would serve as virtually the only cold-reservoir on Earth. It would be interesting to study the lopsided currents in the ocean and atmosphere in such a case.

This case of polar asymmetry not only *could* be so—it *is* so (to a certain extent) on Earth. The Earth's South Pole is occupied by a nearly circular continent, Antarctica, with the South Pole nearly centered on it. Indeed, all my figures for a polar continent's icecap are based on the actual state of affairs in Antarctica, the ice of which contains 2 per cent of Earth's water supply. (If the Antarctic icecap should ever all melt, it would raise the sea level by about 60 meters, or 200 feet.)

The Earth's North Pole, however, is oceanic and is occupied by a nearly circular arm of the ocean, the Arctic Ocean, which is about as large as Antarctica and is covered with sea ice. Indeed, all my figures for a polar ocean's sea ice are based on the actual state of affairs in the Arctic Ocean.

(The opposite nature of Earth's poles is somewhat spoiled by the fact that there is a miniature Antarctica in the far north, too. The large northern island of Greenland has a huge icecap, too, second in size only to Antarctica's—but with only one tenth the mass of the Antarctica icecap.)

In such a case, it is the south polar region that is Earth's refrigerator, far more than the north polar region is; and the disparity would be greater still were it not for Greenland. It is the cold waters of the ocean off Antarctica that tend to fertilize all the rest of the world. The cold Antarctic waters are rich in oxygen and, being heavy with cold, seep northward at the bottom of the ocean, aerating it. When these cold waters well upward for any reason, they bring minerals up, too, so that where this happens, the ocean teems with life. Without the Antarctic waters, Earth's ocean would support only a comparatively limited amount of life, and Earth's land surface would be the poorer for it, too.

The effect of the Antarctic icebox on the southern half of the planet is enormous. There is an island in the Indian Ocean called Kerguelen Island, after its discoverer, or Desolation Island, after its characteristics. It is a semipolar island, frigid and tempest-ridden, with snow fields and glaciers. It is at a latitude of 49° S. At 49° N, in comparison, we have the cities of Paris, France, and Vancouver, Canada.

I have explained that whether you have a polar ocean or a polar continent, the small falls and rises in summer temperature that are due to Earth's astronomical situation don't produce much change—just a relatively minor expansion and contraction of the sea ice in the case of polar oceans and of the iceberg fields in the case of polar continents.

Ah, but remember that this was on the assumption that Earth's land surface was in the tropics, far from either pole, and this is not so in fact. The land surface is distributed asymmetrically, and that, after the nature of the poles, is the second important asymmetry on the Earth's surface.

As it happens, the Earth's land surface is distributed lopsidedly in favor of the northern hemisphere. This means there is not much land near the South Pole (except for Antarctica itself, of course). In fact, the only continental land area south of 40° S is Patagonia, the narrowing southern tip of South America, and a frigid and unattractive land it is.

Consequently, the southern hemisphere is fairly immune to the effects of Ice Ages or of the mild periods between. The Antarctica icecap has been there very much as it is today for at least 20,000,000 years. A little expansion or contraction of the iceberg fields and that's it.

The polar ocean at the North Pole, however, doesn't fit my original assumptions at all. It is *not* open ocean with no land within thousands of kilometers. The Arctic Ocean is, in fact, almost landlocked, with the only considerable connection with the rest of the ocean a 1,600-kilometer (1,000-mile) -wide stretch of water between Greenland and Scandinavia and even this is partially blocked by the island of Iceland.

Furthermore, the land that surrounds the Arctic Ocean is not inconsiderable in extent. North of 40° N is not only Greenland and a number of other large islands, but almost all of Europe and two thirds of North America and Asia.

This northern land makes all the difference. Whereas the snow that falls during the southern-hemisphere winter falls, for the most part, on Antarctica's ice or on ocean water, the snow that falls during the northern-hemisphere winter falls over vast land areas of North America, Asia, and Europe.

The land surfaces on which the snow falls cool down and retain the snow cover through the winter. This means there is snow cover on millions of square kilometers that were bare in the summer.

Those square kilometers are bare again the next summer, for they lie farther from the poles than do Antarctica and Greenland and there is enough solar heat to melt that thin layer of snow completely.

But now suppose the astronomic changes discussed in the previous chapters produce slightly cooler summers in the northern hemisphere, so there is less than usual melting of the snow in the summer for year after year. There will then be a tendency to have a little snow left all through the summer in places like northern Siberia, northern Scandinavia, and northeastern Canada, places where the earlier warmer summers would have melted them.

That sets off the Ice Age trigger by increasing the reflectivity of Earth's surface and therefore cooling the summers further, reducing the melting further, and making sure that an even larger extent of snow covers the bare land areas of the north throughout the year—which further increases Earth's reflectivity, and so on.

This trigger works beautifully, unlike the situation in the open polar ocean or in the sea surrounding a polar continent, for on land the trigger action is not blunted by oceanic circulation.

The snow piles up, turns to ice, and then to glaciers, which advance southward. At the height of an Ice Age, there are five extensive ice sheets on Earth. There are the two that have been permanent for millions of years (Greenland and Antarctica) and, in addition, three that form only in the Ice Ages and only in the northern hemisphere. These are the Canadian, the Scandinavian, and the Siberian, with the Canadian the largest.

All in all, when the ice sheets are at their farthest extent, they cover more than 45,000,000 square kilometers (17,400,000 square miles) of land, three times the amount covered by the present two icecaps. This amounts to just about one third of the Earth's land surface. The total volume of ice may amount to 75,000,000 cubic kilometers (18,000,000 cubic miles), which amounts to about one twentieth of the total water supply on Earth.

Amazingly, it turns out that even at the maximum depth of an Ice Age, the ocean remains little affected. The amount of water in the ocean, when the ice has spread out as far as it ever gets, is 97 per cent of what it is now, when two of the ice sheets still exist. Sea life is scarcely affected, except perhaps for the better, as the ocean cools slightly and holds more oxygen.

What's more, the accumulation of ice on land surfaces lowers the sea level by about 100 meters (320 feet) so that the continental shelves are virtually exposed, this new land making up for the land covered by the ice.

Then, of course, when the astronomical situation shifts back to warmer summers and there is a little more melting in the summer than there is ice formation in the winter, the reverse trigger is set into operation and the glaciers begin their retreat.

This asymmetry of Ice Age situations in the northern and southern hemispheres still doesn't explain why Ice Ages are characterstic of the last million years and not of vast stretches of time before that.

That is because the dramatic and spectacular scenario that involves the formation and disappearance of huge ice sheets depends upon two things, a polar ocean to serve as a water supply and large continental areas crowding in on it to serve as a land base for glacier formation.

That is exactly the situation now, but it was not always so. There is "continental drift," you see, so that the pattern of the continents continually changes. Until a million years ago, apparently, the Arctic Ocean was too open, and the nearest continental areas were too far south to serve as adequate bases for ice sheets. Even slightly cooler summers didn't get cool enough that far south to allow the ice to accumulate.

Apparently the astronomical movements described in the two previous chapters only produce dramatic results when either pole gets land-crowded, without quite being land-occupied, as the North Pole is now—and that seems to happen only every 250,000,000 years or so.

It's my guess that the asymmetry of continental arrangements has allowed a succession of Ice Ages only during 1 per cent of the Earth's history (assuming Earth's astronomical situation to have always been what it is today), and we just happen to have had the human species developing at a time that is at the short end of that 1-in-a-100 chance.

# V

# OUR SOLAR
# SYSTEM

## Ten
# THE COMET THAT WASN'T

I have just received a phone call from a young woman who asked to speak to me about one of my books.

"Certainly," I said. And then, with sudden alarm at her tone, I asked, "Are you weeping?"

"Yes, I am," she said. "It's not really your fault, I suppose, but your book made me feel so sad."

I was astonished. My stories, while excellent, are chiefly noted for their cerebral atmosphere and tone and are not usually considered remarkable for their emotional content. Still, one or two of my stories might pluck at the heartstrings,* and there's something a little flattering about having your writing reduce someone to tears.

"Which book are you referring to, miss?" I asked.

"Your book about the Universe," she said.

If I had been astonished before that was nothing compared to my confusion now. *The Universe* (Walker, 1966) is a perfectly respectable volume, written in a logical and sprightly manner, and doesn't possess one word calculated to elicit tears. Or so I thought.

I said, "How could that book make you feel sad?"

"I was reading about the development of the Universe and about how it must come to an end. It just made me feel there was no *use* to anything. I just didn't want to live."

I said, "But, young woman, didn't you notice that I said our Sun had at least eight billion years to live and that the Universe might last hundreds of billions of years?"

"But that's not forever," she said. "Doesn't it make *you* despair? Doesn't it make astronomers just not want to live?"

"No, it doesn't," I said, earnestly. "And you mustn't feel

* "The Ugly Little Boy," for instance, which you can find in my book *Nine Tomorrows* (Doubleday, 1959).

that way, either. Each of us has to die in much less than billions of years, and we come to terms with that, don't we?"

"That's not the same thing. When we die, others will follow us, but when the Universe dies there's nothing left."

Desperate to cheer her, I said, "Well, look, it may be that the Universe oscillates and that new universes are born when old ones die. It may even be that human beings may learn how to survive the death of a Universe in time to come."

The sobbing seemed to have diminished by the time I dared let her hang up.

For a while, I just sat there staring at the telephone. I am myself a notoriously soft-hearted person and cry at movie listings, but I must admit it would never occur to me to cry over the end of the Universe billions of years hence. In fact, I wrote about the end of the Universe in my story, "The Last Question"† and was very upbeat about it.

Yet as I sat there, I felt myself beginning to think that astronomy might be a dangerous subject and one from which sensitive young women ought to be shielded. Surely, I thought, I can't let myself fall into *that* trap, so the only thing I can do now is to sit down immediately at my typewriter and determinedly begin an astronomical essay.

Let's begin with the number seven, a notoriously lucky number. It is used in all sorts of connotations that make it seem like the natural number for important groups. There are the seven virtues, the seven deadly sins, the seven wonders of the world, and so on, and so on.

What makes seven so wonderful?

You could decide that it is because of some numerical property. Perhaps we might feel that there was something wonderful about its being the sum of the first odd number and the first square; or that there is something about the fact that it is the largest prime under ten that is significant.

I don't think so. I suspect that seven was lucky long before people grew sophisticated enough to become mystical about numbers.

My own feeling is that we have to go back in time to a point where there were seven objects that were clearly exactly seven, clearly important, and even clearly awe-

† It is also in *Nine Tomorrows*.

inspiring. The impressive nature of those objects would then cast an aura of holiness or good fortune on the number itself.

Can there be any question that the objects I'm referring to must be the traditional seven planets of ancient times, the objects which we now call Sun, Moon, Mercury, Venus, Mars, Jupiter, and Saturn?

It was the ancient Sumerians, some time in the third millennium B.C., who made the first systematic observations of these seven bodies and observed the manner in which each changed position from night to night relative to the fixed stars.‡

The changing patterns of the planets against the constellations through which they passed in their more or less complicated movements* were gradually assumed to have significance with respect to Earthly affairs. Their influence in this respect was more than human power could account for, and they were naturally considered gods. The Sumerians named the planets for various gods in their pantheon, and this habit has never been broken in Western history. The names were changed, but only to those of other gods, and at this very time, *we* call the planets by the names of Roman gods.

It was from the seven planets that the custom of the seven-day period we call the week arose in Sumeria, with each day presided over by a different one of them, and that is reflected in the names of those days.†

The Jews picked up the notion of the week during the Babylonian Captivity but devised a Creation story that accounted for the seven days without reference to the seven planets—since planet-gods were not permitted in the strict monotheism of postexilic Judaism.

But if the number seven lost the holiness of the planets in the Judeo-Christian ethic, it gained the holiness of the Sabbath. The aura of inviolability seems, therefore, to have persisted about the seven planets. It was somehow unthinkable that there should be eight, for instance, and

‡ It is this position change which gave rise to the word "planet," for that is from the Greek for "wandering."
* See "The Stars in Their Courses," in the book of that name (Doubleday, 1971).
† See "Moon over Babylon," in *The Tragedy of the Moon* (Doubleday, 1973).

that feeling persisted through the first two centuries of modern science.

After the Polish astronomer Copernicus presented his heliocentric theory in 1543, the term "planet" came to be used for only those bodies that moved about the Sun. Mercury, Venus, Mars, Jupiter, and Saturn were still planets under the new dispensation, but the Sun itself was not, of course. Nor was the Moon, which came to be called a "satellite," a name given to those bodies that circled primarily about a planet, as the Moon circled about the Earth. To counterbalance the loss of the Sun and Moon, the Earth itself came to be considered a planet in the Copernican theory.

Still, that was just nomenclature. Whatever one called the various wandering bodies in the sky visible to the unaided eye, there were exactly seven of them, and we shall still refer to them as the "seven traditional planets."

In 1609 the Pisan astronomer Galileo turned his telescope on the sky and discovered that there were myriads of fixed stars that were too faint to be seen by the unaided eye, but which existed just the same. Despite this, no one seems to have suggested that, in analogy, new planets might also be discovered. The inviolability of the traditional, and sacred, number seven seemed firm.

To be sure, there were also bodies, unseen by the unaided eye, in the solar system itself, for in 1610 Galileo discovered four smaller bodies circling Jupiter, satellites to that planet as Moon is satellite to Earth. Then, before the century was over, five satellites of Saturn were discovered, making a total of ten satellites in all that were known, when our own Moon is included.

Nevertheless, even that didn't alter the sacred number of seven. By defiant illogic, our Moon retained its separate place, while the satellites of Jupiter and Saturn were lumped with the respective planets they circle. We can rationalize this by saying that there are still only seven *visible* wandering bodies in the sky—visible to the unaided eye, that is.

There were the comets, of course, which wandered among the stars, too, but their appearance was so atypical and their comings and goings so unpredictable that they didn't count. Aristotle felt that they were atmospheric exhalations and part of the Earth rather than of the sky.

Others suspected they were special creations, sent across the sky as one-shots, so to speak, in order to foretell catastrophe.

Even in 1758, when the English Astronomer Royal Edmund Halley's prediction that the comet of 1682 (now called "Halley's comet" in his honor) would return in that year was verified and it was understood that comets moved in fixed orbits about the Sun, they were *still* not included among the planets. The appearance remained too atypical, and the cigar-shaped orbits too elongated for them to be allowed on the sacred precincts. —

And yet the odd thing is that there *is* an additional wanderer that fulfills all the criteria of the traditional seven. It is visible to the unaided eye, and it moves relative to the fixed stars. It cannot be denied the right to be considered an additional planet, so just for a while let us call it "Additional."

Why was Additional never observed for all the centuries down to the eighteenth? To answer that, let's ask why the seven traditional planets *were* observed.

For one thing, they are bright. The Sun is the brightest object in the sky by far, and the Moon, thought a very poor second, is second. Even the remaining five traditional planets, which are starlike points far dimmer than the Sun and the Moon, are nevertheless brighter than almost anything else in the sky. In Table 14 the magnitude of the seven planets is given, along with that of Sirius and Canopus, the two brightest of the fixed stars—and Additional. (I discuss magnitudes, by the way, in Chapter 13.)

TABLE 14

| Object | Magnitude and brightest | Brightness (Sirius = 1) |
|---|---|---|
| Sun | −26.9 | 15,000,000 |
| Moon | −12.6 | 30,000 |
| Venus | − 4.3 | 14 |
| Mars | − 2.8 | 3.5 |
| Jupiter | − 2.5 | 2.5 |
| Sirius | − 1.4 | 1.0 |
| Mercury | − 1.2 | 0.9 |
| Canopus | − 0.7 | 0.5 |
| Saturn | − 0.4 | 0.4 |
| Additional | + 5.7 | 0.0015 |

As you see, the five brightest of the traditional planets are also the five brightest objects in the sky. Even the two dimmest of the traditional planets are not far behind Sirius and Canopus. So it is clear that the seven traditional planets attract the eye, and anyone observing the sky in primitive times would see them even if he saw very little else.

Additional, on the other hand, is only 1/700 as bright as Sirius and only 1/270 as bright as Saturn. While it is visible to the unaided eye, it is just *barely* visible.

Of course, brightness isn't the only criterion. Sirius and Canopus are of planetary brightness, but no one ever mistook them for planets. A planet had to shift its position among the fixed stars, and the faster it shifted, the more readily it was noticed.

The Moon, for instance, shifts most rapidly—by an average of 48,100 seconds of arc per day, a distance which is nearly twenty-six times its own width. If one were to watch the Moon at night for a single hour under Sumerian conditions (clear skies and no city lights) that would be enough to show the shift unmistakably.

The rest of the planets move more slowly, and in Table 15 the average shift per day is given for each of them, with Additional included.

Of the seven traditional planets, you see that Jupiter and Saturn are the slow-shifting ones, with Saturn by far the slower of the two. It takes Saturn 29.5 years to accumulate shift enough to circle the entire sky. For that reason, Saturn may have been the last planet to have been recognized in old days, since it was both the least bright and the least fast. (Mercury, which competes for that honor, is in some ways the hardest to see since it is always near the Sun, but

### TABLE 15

| Planet | Average shift (seconds of arc per day) | Days to move the width of the Moon |
|--------|----------------------------------------|-------------------------------------|
| Moon | 48,100 | 0.038 |
| Mercury | 14,900 | 0.125 |
| Venus | 5,840 | 0.319 |
| Sun | 3,550 | 0.525 |
| Mars | 1,910 | 0.976 |
| Jupiter | 302 | 6.17 |
| Saturn | 122 | 15.3 |
| Additional | 42.9 | 43.5 |

once it is glimpsed at sunset or at dawn, its extraordinarily rapid motion may give it away at once.)

But what about Additional, which is only 1/270 as bright at Saturn and which shifts at only a little over ⅓ its speed? That combination of dimness and slowness is fatal. No observer in ancient times and very few even in early telescopic times were likely to look at that object from night to night. There was nothing that made it seem more remarkable than any of the remaining two or three thousand stars of the same brightness. Even if astronomers did actually look at it for a few nights in a row, its slow motion was not likely to make itself overwhelmingly obvious.

So Additional went unnoticed—at least as a planet. Anyone with 20/20 vision who looked in its direction would see it as a "star," of course, and anyone with a telescope certainly would.

In fact, an occasional astronomer with a telescope, plotting the position of the various stars in the sky, might have seen Additional, have plotted it as a star, and even given it a name. In 1690 the first Astronomer Royal, John Flamsteed, noted it in the constellation Taurus, recorded it, and called it "34 Tauri."

Afterward, some other astronomer might have seen Additional in a different place, plotted its new position, and even given it a different name. There would have been no reason to identify the new star with the old star. In fact, the same astronomer might have recorded it in slightly different positions on different nights—each time as a different star. The French astronomer Pierre Charles Lemonnier apparently recorded the position of Additional thirteen different times in thirteen different places in the middle 1700s, under the impression that he was recording thirteen different stars.

How was this possible? Two reasons.

The other planets were, first of all, clearly planets, even if one disregarded their motion and their brightness. Planets were not points of light as the stars were; they were round discs. The Sun and Moon appeared as discs to the unaided eye, while Mercury, Venus, Mars, Jupiter, and Saturn all appeared as discs even through the primitive telescopes of the seventeenth and eighteenth centuries. Additional, however, did not show up as a disc in the telescopes of men like Flamsteed and Lemonnier, and in the absence of a disc, why should they think in terms of planets?

And the second reason is that the sevenness of the traditional planets was so well entrenched in the common thinking of man, that Additional, as a planet, was unthinkable, and so astronomers didn't think of it. You might as well suddenly decide you had discovered an eighth day of the week.

But then upon the scene came Friedrich Wilhelm Herschel, born in Hannover on November 15, 1738. Hannover was a then-independent state in what is now West Germany, and for historical reasons its ruler happened to be King George II of Great Britain.

Herschel's father was a musician in the Hannoverian army and Herschel himself entered the same profession. In 1756, however, the Seven Years' War began (an odd coincidence that the number seven should figure crucially in Herschel's life in so completely nonplanetary a way), and the French, fighting Prussia and Great Britain, occupied the Hannoverian realm of the British monarch in 1757.

The young Herschel, unwilling to suffer the miseries of an enemy occupation, managed to wriggle out of Hannover, deserting the army in the process, and got to Great Britain, where he remained the rest of his life and where he Anglicized his Christian names to a simple "William."

He continued his musical career and by 1766, he was a well-known organist and music teacher at the resort city of Bath, tutoring up to thirty-five pupils a week.

Prosperity gave him a chance to gratify his fervent desire for learning. He taught himself Latin and Italian. The theory of musical sounds led him to mathematics and that, in turn, led him to optics. He read a book that dealt with Isaac Newton's discoveries in optics and he became filled with a fervent and lifelong desire to observe the heavens.

But for that he needed a telescope. He couldn't afford to buy one, and when he tried renting one, it turned out that its quality was poor and he was very disappointed at what he saw—or, rather, didn't see.

He came to the decision at last that there was nothing to do but to attempt to make his own telescopes and, in particular, to grind his own lenses and mirrors. He ground two hundred pieces of glass and metal without making anything that satisfied him.

Then, in 1772, he went back to Hannover to get his sister, Caroline, who, for the rest of her life, assisted first William, then his son, John, in their astronomic labors with a single-minded intensity that precluded marriage or virtually any life for herself at all.‡

With Caroline's help, Herschel had better luck. While he ground for hours at a time, Caroline would read to him and feed him. Eventually, he got the trick of grinding and developed telescopes good enough to satisfy him. In fact, the musician who could not afford to buy a telescope ended by making for himself the best telescopes then in existence.

His first satisfactory telescope, completed in 1774, was a 6-inch reflector, and with it he could see the Great Nebula in Orion and clearly make out the rings of Saturn. That was not bad for an amateur.

Much more was ahead, however. He began to use his telescope systematically, passing it from one object in the sky to another. He bombarded learned bodies with papers on the mountains on the Moon, on sunspots, on variable stars, and on the Martian poles. He was the first to note that Mars's axis was tilted to its plane of revolution at about the same angle as Earth's was, so that the Martian seasons were essentially like Earth's, except that they were twice as long and considerably colder.

Then, on the night of Tuesday, March 13, 1781, Herschel, in his methodical progress across the sky, suddenly found himself looking at Additional.

There was now an important difference. Herschel was looking at Additional with a telescope that was far superior to any of those used by earlier astronomers. Herschel's telescope magnified the object to the point where it appeared as a *disc*. Herschel, in other words, was looking at a disc where no disc was supposed to be.

Did Herschel jump at once to the notion that he had found a planet? Of course not! An additional planet was unthinkable. He accepted the only possible alternative and announced that he had discovered a comet.

‡ She did make astronomic observations of her own eventually, with a telescope William made for her. She discovered eight comets, was the first woman astronomer of note, and died, at last, just ten weeks short of her ninety-eighth birhday.

But he kept on observing Additional and by March 19 could see that it was shifting position with respect to the fixed stars at a speed only about a third as great as that of Saturn's shift.

That was a troublesome thing. Ever since ancient Greek times, it had been accepted that the more slowly a planet shifted against the stars, the farther it was likely to be from us, and the new telescopic astronomy had confirmed that, with the modification that it was distance from the Sun that counted.

Since Additional was shifting much more slowly than Saturn, it had to be more distant from the Sun than Saturn was. Of course, comets moved in orbits that took them far beyond Saturn, but no comet could be seen out there. Comets had to be much closer to the Sun than Saturn was in order to become visible.

What's more, Additional's motion was clearly in such a direction as to indicate the object was making its way through the signs of the zodiac, as all the planets did, but as virtually none of the comets did.

Then, on April 6, 1781, he managed to get a good enough view of Additional to see that the little disc had sharp edges like a planet, not hazy ones, like a comet. Wha's more, it showed no signs of a tail.

Finally, when he had enough observations to calculate an orbit, he found that orbit to be nearly circular like that of a planet, and not very elongated like that of a comet.

Reluctantly, he had to accept the unthinkable. His comet wasn't; it was a planet. What's more, from its slow shift, it lay far beyond Saturn; it was just twice as far from the Sun as Saturn was.

At one bound, the diameter of the known planetary system was doubled. From 2,850,000,000 kilometers (1,770,000,000 miles), the diameter of Saturn's orbit, it had risen to 5,710,000,000 kilometers (3,570,000,000 miles), the diameter of Additional's orbit. It is Additional's great distance that is responsible for its dimness, its slow shift against the stars, its unusually small disc—in short, for its very belated recognition as a planet.

Now it was up to Herschel to name the planet. In a bit of excess diplomacy, he named it after the then-reigning sovereign of Great Britain, George III, and called it "Georgium Sidus" ("George's star"), an uncommonly poor name for a planet.

King George was, of course, flattered. He officially pardoned Herschel's youthful desertion from the Hannoverian army and appointed him his private court astronomer at a salary of three hundred guineas a year. As the discoverer of a new planet, the first new one in at least five thousand years, he at once became the most famous astronomer of the world, a position he retained (and *deserved*, for he made many other important discoveries) to the end of his life. Perhaps most comforting of all, he married a rich widow in 1788, and his financial problems were nonexistent thereafter.

Fortunately, for all Herschel's new-found prestige, the name he gave Additional was not accepted by the indignant intellectuals of Europe. They weren't going to abandon the time-honored practice of naming planets for the classical gods just in order to flatter a British king. When some British astronomers suggested "Herschel" as the name for the planet, that was rejected, too.

It was German astronomer Johann Elert Bode who offered a classical solution. The planets that are farther from the Sun than Earth is, present a sequence of generations. Those planets, in order, are Mars, Jupiter, and Saturn. In the Greek mythology Ares (the Roman Mars) was the son of Zeus (the Roman Jupiter), who was the son of Kronos (the Roman Saturn). For a planet beyond Saturn, it is only necessary to remember that Kronos was the son of Ouranos (the Roman Uranus). Why not, then, call the new planet "Uranus"?

The notion was accepted with a glad cry, and Uranus it was, and has remained ever since.

Oddly enough, the sacred seven was not really disturbed by the discovery of Uranus. Rather, it was restored! By the Copernican system, in which the Sun and Moon are *not* planets and Earth *is*, there were just six known planets prior to 1781. These, in order of increasing distance from the Sun, were Mercury, Venus, Earth, Mars, Jupiter, and Saturn. Once Uranus was added the number of Copernican planets became *seven!*

As Herschel's reputation and wealth grew, he built ever bigger and better telescopes. He returned to his planet Uranus in 1787 and found two satellites circling it, the eleventh and twelfth known to exist (counting our

Moon).* These satellites were eventually named Titania and Oberon, after the queen and king of the fairies in Shakespeare's *A Midsummer Night's Dream*. It was the first time that classical mythology had been abandoned in naming the satellites.

These satellites introduced an interesting anomaly—

The axes of several of the planets were tipped from the perpendicular to the plane of their orbital revolutions. Thus, Saturn's axis was tipped 27 degrees, Mars' was tipped 24 degrees, and Earth's 23.5 degrees. Jupiter's axis was a little unusual in being tipped only 3 degrees.

The planes of the orbital revolutions of the satellites of Jupiter and Saturn were tipped to the same extent that the axes of those planets were. The satellites revolved in the plane of the planetary equator.†

But the satellites of Uranus moved in a plane that was tipped 98 degrees from the perpendicular to the plane of Uranus' orbit. Could it be that Uranus' axis was tipped by that much and was very nearly in the plane of its orbital revolution? If so, Uranus would seem to be lying on its side, so to speak, as it moved around the Sun.

This extreme axial tip was eventually confirmed and to this day, astronomers have no adequate explanation as to why Uranus, alone of all the known planets, should be lying on its side.

And yet this is not the most dramatic thing to have come out of the study of Uranus—and that I will discuss in Chapter 11.

* In 1789 he discovered two more satellites of Saturn, making seven for that planet and fourteen altogether.

† See "The Wrong Turning," in *The Planet That Wasn't* (Doubleday, 1976).

## Eleven
# THE SEA-GREEN PLANET

I had arrived at a certain university some time ago in order to give a talk, and I was being guided about the place by a young woman who was a student at the institution.

She was approaching a swinging door in advance of me when a tall young man (presumably another student) cut her off, went through, and allowed the door to swing back violently, catching her in the chest and sending her staggering backward.

Rather annoyed, I went through the swinging door at a run and called after him loudly, "Well done, big shot. Good work."

He stopped cold and slowly turned, his eyebrows coming together in a frown and his lips twisted in a scowl. Perhaps he didn't know to what I was referring, but he must have gathered from my expression that I disapproved of him.

He approached menancingly and, since I didn't know what else to do, I stood my ground.

He said, "Something eating you, mister?"

"Not really," I said. "It's just that you went through that door and let it swing back and knock a girl over, and I wanted to congratulate you on your marksmanship."

Apparently, the young man was not accustomed to sarcasm delivered in a pleasant tone of voice. He brooded about it, groped for some phrase from his limited armory, and said, "Watch your (expletive deleted) language, mister."

"Very well," I said. "Of which of the words that I used do you disapprove?"

That stopped him again, so he reached for another phrase. "I don't like your manner, mister."

There he stood, six inches taller than myself and considerably less than half my age. Hoping fervently that my gray hairs would protect me, I smiled and said, "And what do you plan to do about it?"

Actually, I was very nervous concerning what plans he might have but, to my relief, he said, "It so happens I don't believe in physical violence."

"Good!" I said. "Then why did you knock down the young woman?"

"That was an accident," he said.

"I haven't heard you apologize."

He looked at me, he looked at the young woman (who was even more fearful of my being broken in two than I was, since she was in charge of me), and then, unable to think of any alternative to apology but flight, turned and strode away.

It was very distressing. I cling rather wistfully to the theory that everyone is nice, so that nasty people tend to upset my Universe. And yet, when I'm being rational about it, I *know* that there are nasty people here and there on Earth, even among scientists.

Consider the case of Nice Guy Adams and Nasty Guys Challis and Airy—

When Isaac Newton had worked out his theory of gravitation, the equation he produced applied to a situation in which only two bodies were involved. If the Moon and the Earth were the only bodies in the Universe, Newton's equation would describe the path of the Moon and the Earth about their common center of gravity with great precision. The "two-body problem" is solved.

As soon as you deal with three bodies—say, the Moon, the Earth, and the Sun—their motions cannot be expressed *exactly* by Newton's equation or by any equation developed since. The "three-body problem" is still not solved.

Actually, that doesn't matter, except to the theoreticians. In actual fact, even though the Universe does not contain merely three bodies, but countless trillions of them, Newton's equation still works well enough.

If you want to describe the Moon's motion around the Earth, you first work it out as though only the Moon and the Earth exist. That gives you a very good first approximation.

You then calculate the much smaller effects of more distant bodies. Since the gravitational pull between two bodies falls off as the square of the distance between their centers and since all other astronomical bodies are much farther from the Earth and the Moon than those two bodies are from each other, you expect the other effects to be small.

The Sun, however, is so large that even at its great distance its gravitational pull is significant. As the Moon moves in its orbit about the Earth, it is sometimes a little closer to the Sun than Earth is, sometimes a little farther than Earth. Tht two bodies feel the Sun's gravitational pull to slightly and changeably different degrees. This introduces a small modifying effect on the Moon's motion that can be calculated for.

The still smaller pull of Venus, varying with the distance of that planet from the Earth and the Moon can also be allowed for. So can the pull of Mars, the pull of Jupiter, and so on.

To allow for all these pulls on the Moon, in all their variations with time, produces an approximate equation (never an exact one) that is so enormously complicated that Newton said the problem of the Moon's motion was the only one that ever made his head ache.

These various minor pulls, which cause an orbital motion to vary somewhat from what it would be if the two neighboring bodies were the only ones in existence, are called "perturbations."

Theoretically, every object in the Universe can produce a perturbation affecting the motion of every other body. In practice, the more massive the body being perturbed, the less massive the body doing the perturbing, and the greater the distance between the two—the smaller the perturbation. The perturbing effect of an Earth-launched planetary probe on the planet it passes or the perturbing effect of the star Aldebaran on the Moon are immeasurably small and can be dismissed.

Using Newton's equation and allowing for all perturbations of reasonable size, the motion of the various planets and satellites of the solar system could be worked out with reasonable precision. From Mercury to Saturn all the worlds marched across the sky just about as the equation would predict. The astronomers of the early decades of the 1800s had instruments that could make close measurements; Newton's equation fitted within those measurements, and the astronomers were happy.

But what about Uranus? That was a planet that had not been known in Newton's time, that had only been discovered in 1781, as I described in the previous chapter. Would Newton's equation fit there?

It seemed to be a rather simple case, since Uranus, far out on the edge of the solar system seemed far away from any perturbing influences. The known body closest to Uranus was Saturn, which, at its closest, was 1,500,000,000 kilometers (930,000,000 miles) from it. The known body next closest to Uranus was Jupiter, which, at its closest, was 2,100,000,000 kilometers (1,300,000,000 miles) from it.

This meant that in calculating the orbit of Uranus about the Sun, one had to allow for a small perturbing effect from Saturn and a small perturbing effect from Jupiter, and that was all. All other bodies in the Universe, as far as was known, were too small or too distant, or both too small and too distant, to produce perceptible perturbations.

So Uranus' motion across the sky was watched with interest and its position compared to theory from year to year.

But then came trouble. In 1821 the French astronomer Alexis Bouvard collected all the observations of Uranus deliberately made since its discovery and accidentally made before its discovery when it was entered into the star charts, occasionally, as a star. He tried to fit this to the calculated orbit of Uranus and it wouldn't fit. He recalculated the perturbing effect of Jupiter and Saturn with great care, and still the actual position of the planet refused to coincide with the calculated position.

The difference between the actual position Uranus occupied and the theoretical position it was supposed to occupy was never very great—not more than 2 minutes of arc, or one fifteenth the apparent diameter of the Moon —but you didn't pay off on "never very great." What astronomers wanted was "vanishingly small."

How, then, to account for Uranus' behavior?

One possible explanation was that Newton's equation was slightly wrong. According to that equation, the strength of gravitational pull between two bodies fell off as the square of the distance between their centers. This is the "inverse-square law."

But suppose the gravitational fall-off was not *quite* as the square of the distance. Suppose the factor were not $d^2$ but $d^{2.0001}$ or $d^{1.9999}$. In that case there would be a discrepancy between the calculated motion obtained through use of the inverse-square law and the real motion that depended on the slightly different law. What's more, the greater the distance between two bodies, the greater the discrepancy.

Out as far as Saturn, the discrepancy, if any, had to be small enough to escape detection, since all the large bodies out to that point followed their calculated paths with precision. At the distance of Uranus from the Sun (twice that of Saturn) the discrepancy might have expanded to the point of detectability. Furthermore, the distances between Uranus and the two perturbing bodies, Saturn and Jupiter, were greater than the equivalent distances for planets closer to the Sun, so that the perturbations might be measurably distorted through use of the inverse-square law, too.

Astronomers were, however, reluctant to tamper with Newton's equation until all other alternatives had been ruled out. One reason for this was aesthetic. The inverse-square law could be so simply represented in mathematical form that it was "elegant" and no scientist likes to disturb elegance until all else fails.

Another reason was practical. If the inverse-square law was modified to account for Uranus' motion, it would be an *ad hoc* adjustment. The Latin phrase *ad hoc* means "for this purpose" and is used for any argument that is brought in solely for the purpose of explaining some one thing that is otherwise puzzling, especially if the argument cannot be applied to any other phenomenon.

Even though an *ad hoc* adjustment of the inverse-square law would fit Uranus, there was no other body in the solar system it could fit since only Uranus was far enough away for the adjustment to be meaningful—and an adjustment for Uranus alone was unconvincing.

Even though an *ad hoc* adjustment of the inverse-sqpare law would fit Uranus, there was no other body in the solar system it could fit since only Uranus was far enough away for the adjustment to be meaningful—and an adjustment for Uranus alone was unconvincing.

Of course, if there were another distant planet, its motion could be tested, too, and if its motions *also* fitted the adjustment of the inverse-square law, then the argument would become rather convincing.

But if another distant planet existed beside Uranus, it might be the source of a perturbing gravitational pull on

Uranus, which would account for the discrepancy in Uranus' motion. You might not, in that case, *need* the adjustment of the inverse-square law.

Some astronomers seized on that possibility. Another planet, meaning another gravitational pull, meaning another perturbation, meaning a corrected orbit that Uranus would follow, would be delightful. —But where was the planet?

It couldn't be closer to the Sun than Uranus was, for then, if it were large enough to produce a perceptible perturbation on Uranus, it should also be large enough to he detected without trouble, and it hadn't been. Furthermore, it should then also produce an unaccounted-for perturbation of Saturn's orbit, and there was none.

The unknown planet, if any, would have to be farther away than Uranus, displaying a smaller disc and a slower motion than any other planet and therefore having escaped detection till then. Further, from a point far enough beyond Uranus, it would be close enough to that planet to perturb it noticeably, but too far from Saturn to perturb that planet noticeably.

It was not enough to postulate a distant Planet Eight beyond Uranus. The planet had to be detected. But if Uranus could barely be seen by the unaided eye, Planet Eight, dimmer still, would surely be visible only by telescope and, with its tiny disc and slow motion, would be lost among the great numbers of equally dim stars. Detection would be difficult indeed.

But why not turn things about? If you know where a planet is and how it moves, you can calculate its perturbing effect on Uranus. Given the perturbing effect, can you not calculate where the planet is and how it moves and therefore know where to look for it?

Enter the British scientist John Couch Adams who, in 1841, was twenty-two years old and who was studying at Cambridge University. He was first in his class in mathematics and it occurred to him to try to calculate the position of Planet Eight.

If Planet Eight were on the other side of the Sun from Uranus as both pursued their slow passage along their vast orbits, the distance between the two planets would be too great for a detectable perturbation of Uranus to exist. Therefore, both must be on the same side of the Sun.

Since Uranus' position was a little in advance of its calculated position, Planet Eight must, during all or most of the years since Uranus' discovery, have been ahead of it, so that its gravitational pull had served to hurry Uranus along a little. Uranus, however, being closer to the Sun, would move faster than Planet Eight and would therefore overtake it. (The overtaking took place in 1822, actually.) Thereafter, Planet Eight would be behind Uranus and would tend to slow Uranus' motion slightly. All this had to be taken into account.

Adams made some simplifying assumptions to begin with. He assumed Planet Eight was about as massive as Uranus, moved in a perfectly circular orbit in the same plane as Uranus, and was twice as far from the Sun as Uranus was (just as Uranus was twice as far from the Sun as Saturn was).

All these assumptions were chosen to make the calculations easier, but they were reasonable ones. Using them and the observed discrepancies of Uranus' position from year to year, Adams worked away in his spare time and, by September 1845, had calculated the position of Planet Eight for October 1 of that year. It was at a point within the constellation of Aquarius.

Naturally, the planet wouldn't be exactly at that point unless all of Adams' assumptions were exactly correct and they clearly might not be. (One of them turned out to be badly off, for Planet Eight was not twice as far from the Sun as Uranus was, but was only 1.5 times as far.) Anyone would know, therefore, that it would not be enough to look at the exact predicted point, but that one would have to sweep the neighborhood of the point and carefully study thousands of stars.

Adams gave the result of his calculations to James Challis (the first villain of the piece) since Challis was the director of the Cambridge Observatory. Adams' hope was that Challis, with telescopes at his disposal, would search Aquarius for the planet. Challis thought otherwise. Knowing quite well the search would be tedious and was more likely than not to end with nothing, he ducked. He gave Adams a letter of recommendation to the Astronomer Royal, George Biddell Airy (the second villain), and thus successfully passed the buck.

Airy was a conceited, envious, small-minded person who ran the Greenwich Observatory like a petty tyrant. He was obsessed with detail and invariably missed the big picture.

Thus, later in his life, he labored over expeditions intended to study the transits of Venus across the Sun in 1874 and in 1882. By determining the exact time at which Venus made apparent contact with the solar disc as seen from different observation points, the distance of Venus from Earth and, therefore, all the other planets and the Sun could be calculated (it was hoped) with unprecedented accuracy.

Airy spent years training his observers, building a model of a Venus transit on which dry runs could be made, personally making sure that everything was packed and labeled in the most meticulous way, getting every last minor detail exactly right, as though his underlings were one and all five years old—and never considering what effect Venus' dense atmosphere might have. As it turned out, its atmosphere obscured the precise moment of contact of Venus and the solar disc and made the entire expedition worthless.

About the only real success Airy had was personal. He was the first to design eyeglass lenses that corrected astigmatism. He was himself astigmatic, you see—literally, as well as figuratively.

It was this nasty person that Adams tried to contact. The telephone had not yet been invented, so Adams traveled twice to Greenwich and twice Airy wasn't home. A third time, Airy was at dinner and would not be disturbed (naturally). Adams left his paper and Airy finally leafed through it and wasn't impressed (naturally). Airy, with his usual flair for picking the wrong solution, was convinced that the inverse-square law needed adjusting and would not accept a new planet. He therefore wasted time by writing to Adams and asking him to check some points that were completely irrelevant to the problem.

Adams knew they were irrelevant, sighed, and gave up. He didn't even answer the letter.

Meanwhile, in France, a young astronomer, Urbain Jean Joseph Leverrier, was also working on the problem. He made the same assumptions Adams did and located Planet Eight in very nearly the same place in Aquarius that Adams did. He completed his work about half a year after Adams and, of course, had no knowledge of what Adams had done.

Leverrier, who had already established a reputation as an astronomer (as Adams had not) was encouraged by his superiors (as Adams was not) and published his calculations.

Airy read Leverrier's paper, then wrote to him asking him the same irrelevant question he had asked Adams—but without telling Leverrier that Adams had already done the work. Unlike Adams, Leverrier answered at once, pointing out that the question was irrelevant.

By now, Airy was reluctantly impressed. Two men had come up with the same solution, and the silliness of his own objection had been pointed out. He therefore wrote to Challis at Cambridge, asking him to inspect the sky in the indicated position and see if he could find a planet.

Challis was no more eager to take up the search now than he had been in the first place. He didn't think it would come to anything and he was far more interested in some trivial computations he was making of cometary orbits. So he didn't hurry. It was three weeks after he got Airy's request before he even started and then he proceeded very slowly indeed.

By September 18, 1846, he had been at it for six weeks, checking thousands of stars in a desultory sort of way, uninterested, unenthusiastic, and failing to check stars observed on another day in order to see whether any of them moved relative to the rest—which would be a sure sign it was a planet.

On September 18, meanwhile, Leverrier, receiving nothing further from Cambridge and feeling, in any case, that the Berlin Observatory was the best in Europe, wrote to Berlin. The director of the Berlin Observatory was willing to check the matter and he asked the German astronomer Johann Gottfried Galle to take care of it.

Galle might have had to go through the same tedious checking that Challis was going through (though undoubtedly with greater industry and care) were it not for a lucky break. The Berlin Observatory had been preparing a careful series of star charts and a twenty-four-year-old astronomer at the observatory, Heinrich Ludwig d'Arrest, suggested to Galle that he ought to see if Aquarius had been charted.

It *had* been, and only half a year before. Galle got the chart and now the matter was simple. He didn't have to look for a visible disc. He didn't have to study anything from day to day just to see if it were moving against the background of the other stars. All he had to do was to see if there was any object in that section of the sky that was not in the same position on the chart.

On the night of September 23, 1846, then, Galle and D'Arrest got to work. Galle was at the telescope, moving methodically over the field, calling out the positions of stars, one by one, while D'Arrest was at the chart, checking those positions. They had been working for but one hour, no longer, when Galle called off the position of an eighth-magnitude star and D'Arrest said, in excitement, "That's not on the chart!"

It was the planet! It was only fifty-two minutes (about 1.5 times the apparent width of the full Moon) away from the predicted point. Naturally, Galle checked on it from night to night, but after a week it was quite certain that it was moving and that it had a disc. Planet Eight had been discovered.

Once the news was announced, Challis hurriedly went over his own observations and found that he had seen Neptune on four different occasions but had never compared positions and so had not known what he had.

Both Airy and Challis had made fools of themselves and they knew it. They had lost credit for a magnificent discovery that they might have made. Neither one, in their wriggling attempts at self-justification, thought to give proper credit to Adams.

The English astronomer John Herschel knew of Adams' work, however, and he took up the cudgels for him. Herschel was the son of the discoverer of Uranus and an important astronomer in his own right, so his word carried weight. No sooner was the discovery of Neptune announced than Herschel wrote a letter to point out that Adams had done the work before Leverrier and had reached the same conclusions.

Naturally, the French were very indignant at what seemed to them to be an attempt by the British to steal the credit, and for quite a while there was a furious and bitter controversy over the matter in which Adams and Leverrier did *not* join. (They later met and became friends.) It had a happy end, though. Nowadays, the credit for the discovery of Planet Eight is shared by the two men, as it should be.

(It turned out, of course, that Galle had not been the first to sight the planet. Back on May 8, 1795, only fourteen years after the discovery of Uranus, the French astronomer Joseph Jérôme de Lalande noted a star whose position he recorded. Two days later he observed it again and was mortified to note that he had made a mistake in the position. He recorded the new position and forgot the matter. Actually, he had made no mistake. The "star" had moved in those two days, because Lalande, without knowing it, had been looking at Planet Eight.)

What was the new planet to be called? French astronomers, stung by the British claim to precedence, pushed hard to lock in French credit by giving the planet the name "Leverrier." They sweetened this suggestion by proposing that Uranus be deprived of its mythical name and be called "Herschel" (as had indeed been originally suggested by British astronomers). It was pointed out that comets were named after their discoverers and that that had set a precedent.

Everyone but the French astronomers, however, set up a wild howl of protest and the proposal was dropped. It was back to mythology.

Planet Eight has a distinct greenish color when seen in the telescope and Leverrier had this in mind, perhaps, when he suggested, quite early in the game, that the new sea-green planet be named after the Roman god of the green sea, Neptune (equivalent to the Greek god Poseidon). That suggestion was adopted.

—And what happened to poor John Couch Adams after all this? As far as I know, he never succumbed to bitterness. He worked away at astronomy and was able to show that the Leonid meteor swarm had an elongated comet-like orbit, thus strengthening the suggestion that much of the interplanetary debris in the inner solar system consisted of the fragments of disintegrated comets.

In 1860 Challis retired as director of the Cambridge University Observatory and Adams was given the post, in what seems to have been a quiet apology for the past. Then, in 1881, Airy, having been Astronomer Royal for forty-five years, retired, and that post, too, was offered to Adams. This Adams refused, feeling too old to take on the responsibility.

Airy and Adams were curiously united in death as well. Airy died on January 2, 1892, at the age of 90.4 years. Adams followed less than three weeks later, on January 21, 1892, at the age of 72.6 years. Galle, however, survived his sighting of Neptune by over six decades, dying on July 10, 1910, at the age of 98.1 years.

# Twelve
# DISCOVERY BY BLINK

I was interviewed the other day on the subject of immortality and maintained, rather insistently, that it was a bad thing. Immortality, I said, was bad for the species because it would halt its evolution; bad for society because the last generation would be made up of the same increasingly dull individuals indefinitely, and bad for the individual because eventually he would prefer death to boredom. In fact, any society of immortals, I said, would merely shift the pattern from death by circumstance to death by will, with perhaps little change in life expectancy after all.

All this, I think, was not what the interviewer wanted to hear. He therefore personalized the matter and said, "Do you think *you* would want to die someday, assuming you were in good health and could live forever if you chose?"

"Certainly," I said, staunchly.

"When?"

"When I no longer felt the desire to write," I said.

"And when would that be?" he said.

"Never," I said—and killed my whole argument.

Another interviewer once tried to break down my stubborn resistance to any way of spending my life other than at the typewriter by saying to me, "But suppose you knew you had only six months to live. What would you do then?"

And without hesitation, I said, "Type faster."

Well, what's wrong with that attitude? There are many people who are, or were, monomaniacally interested in whatever field of endeavor absorbed them. It's just that most of these fields are not as noticeable to the general public as writing is.

Suppose my mania involved the search for a new and as-yet-undetected planet? Who would see the existence of my madness except a few other astronomers?

—Which, of course, brings me to the subject of planetary discoveries, which this chapter concludes.

In 1781 Uranus, the seventh planet of the solar system, in order of increasing distance from the Sun, was discovered (see Chapter 10) and in 1846 Neptune, the eighth planet, was discovered (see Chapter 11). Was that the end?

No. Uranus had been an accident, but Neptune was glory and triumph, and no astronomer could resist the temptation to repeat that. Astronomers *wanted* more planets to exist.

And why not? The gravitational field of the Sun dominated space, without significant interference from even the nearest stars, for a distance at least a thousand times the distance of Neptune. Across that distance, even if we assumed each planet to be roughly twice as far from the Sun as the one before, there would be room for at least ten trans-Neptunian planets.

Of course, even if those trans-Neptunian planets existed, discovering them would be extraordinarily difficult.

For one thing, the farther from the Sun a planet is, the less light it catches and reflects and the less of that reflected light we ourselves would intercept. Thus, Saturn, at a distance of 1,400,000,00 kilometers (860,000,000 miles) from the Sun, shines brilliantly in our sky with a magnitude of −0.4 and is brighter than all but the two brightest stars.

Uranus, the next planet out, at a distance of 2,800,000,-000 kilometers (1,750,000,000 miles), has a magnitude of only 5.7, which makes it just barely visible to the unaided eye. Neptune, at a distance of 4,500,000,000 kilometers (2,800,000,000 miles) has a magnitude of 7.6 and can never be seen by the unaided eye, but can be seen with a small telescope.

The next planet beyond Neptune would have a magnitude of perhaps 12 or 13 at most and would be seen only with a large telescope. And those still farther out might well be too dim to see with even the largest telescope at our disposal.

Still, astronomers can make out stars with magnitudes siderably dimmer than 12 or 13. Leaving the still more distant planets out of account, there seemed no reason to suppose, as the nineteenth century wore to its end, that the ninth planet, the nearest of the trans-Neptunian planets (assuming it existed), could not be seen.

But even if it could, seeing wasn't enough. The dimmer the object you try to see in a telescope, the greater the number of stars of equal or greater brightness you will also see. Uranus has relatively few stars surrounding it, in a telescopic view, that are as bright as or brighter than itself. Neptune, which is much dimmer, is surrounded by many more stars that successfully compete, and any trans-Neptunian planet would be lost in a veritable snowdrift of stars.

Could the trans-Neptunian planet hidden among the star-powdering be identified? It would have two properties that would at once distinguish it as a planet: it would show a disc, which stars do not, and it would show motion relative to the nearby stars.

The trouble is that the farther away a planet is, the less likely it is to show a perceptible disc. This is especially so if the planets beyond Neptune tend to grow smaller with distance, as can be reasonably argued they do. And as for motion, the farther away a planet is, the slower that motion. In the case of the trans-Neptunian planet, then, you would be dealing with a particularly small disc and a particularly slow motion. Detection would be difficult.

One way of increasing the rather rotten odds against finding the planet would be to try to figure out, at least roughly, where in the sky it might be and then concentrate our looking in that region.

Neptune was discovered because Uranus' orbit indicated the presence of a gravitational pull from beyond. From the nature of the effect of that pull on Uranus' motion, a rough idea was obtained as to the position and distance of Neptune, which was the source of that pull. Neptune was looked for in the indicated place, in the constellation Aquarius, and was found.

Could this process not be repeated? Could not the imperfections in Neptune's orbit be used to locate the ninth planet and then its orbital imperfections used to locate the tenth planet, and so on?

There's a catch. The farther the planet, the longer it takes to complete one revolution about the Sun. The precision with which one can detect imperfections in the orbital movement depends upon the fraction of the turn it has completed.

Thus, Uranus circles the Sun in eighty-four years, and in 1846, when Neptune was discovered, Uranus had been under continuous observation for sixty-five years, or for 0.77 of its revolutionary period. Neptune circled the Sun in one hundred sixty-five years, and in 1900, when it had been under continuous observation for fifty-three years, it had completed only 0.32 of its revolutionary period.

So, as the twentieth century opened, Neptune's orbit was not yet known with sufficient precision for it to be very helpful in locating the trans-Neptunian planet.

Well, then, what about Uranus? By 1900 it had been continuously observed for 1.4 of its revolutionary periods. Once Neptune's pull was taken into account, were there no discrepancies left in Uranus' orbital motion? If the trans-Neptunian planet existed, it should have some effect on Uranus, though a much smaller one than Neptune's effect, since the trans-Neptunian planet would be considerably farther from Uranus than Neptune was.

And, as it happened, Neptune's pull only took care of about 59/60 of the discrepancy that had existed in the orbital calculations of Uranus. There still remained 1/60 unaccounted for, and that must be due to a trans-Neptunian planet. But that was a very small quantity to work with.

There remained other objects existing in the outer regions of the solar system and those were the comets. By the late nineteenth century a number of comets were known whose orbits were calculated. Some of them had aphelia (that is, farthest points from the Sun) in the neighborhood of Jupiter's orbit. It was felt that Jupiter's gravitational pull had fixed the cometary orbits there and those comets were known as the "Jupiter family."

There were comets with aphelia well beyond Jupiter which might conceivably have been affected by the farther planets. In particular, there were several comets (Halley's comet among them) with aphelia well beyond Neptune's orbit, and might they not have been captured by a trans-Neptunian planet?

These were not promising lines of attack—the very small orbital discrepancy of Uranus, the very vague orbital discrepancies of Neptune, and the very uncertain testimony of cometary aphelia—but it would have to do. By about 1900 astronomers were beginning to offer speculations as to the possible orbit of a trans-Neptunian planet.

The gravitational effect of such a planet on Uranus and Neptune would fit best if the source were imagined as moving about the Sun with a particular velocity. Such a velocity would automatically fix the distance of the planet from the Sun. With the distance known, the mass of the planet required to produce its effect on Uranus and Neptune could be calculated. Then, if no circular orbit would fit the facts, the orbit might be imagined as distinctly elliptical and with an orbital plane tipped by a certain amount, so that the distance of the planet from Uranus and Neptune differed considerably from one end of its orbit to the other.

The data which astronomers had, to begin with, was so small, vague, or uncertain that widely different solutions were almost equally possible. One astronomer suggested that the trans-Neptunian planet was more massive than Jupiter and was at a distance of 15,000,000,000 kilometers (10,000,000,000 miles) or over three times the distance of Neptune from the Sun. Others suggested a smaller planet only 6,000,000,000 kilometers (3,700,000,000 miles) from the Sun or less than one and a half times the distance of Neptune. Some suggested two or even three trans-Neptunian planets between the limits of 6,000,000,000 and 15,000,000,000 kilometers from the Sun.

The two most careful calculations, however, were those of the American astronomers Percival Lowell and William Henry Pickering. Both had been born in Boston, Massachusetts, Lowell on March 13, 1855 (the seventy-fourth anniversary of the discovery of Uranus) and Pickering on February 15, 1858.

They were rivals in a way. Lowell was the great proponent of canals on Mars (see "The Olympian Snows," in *The Planet That Wasn't*, Doubleday, 1976) but remained a minority figure among professional astronomers in this respect. Few other observers could see the canals (which were, it now appears, an optical illusion) except occasionally and uncertainly, and absolutely none could see them as clearly and in such detail as Lowell could.

Pickering was the leader of the anti-canal group. He was almost as assiduous as Lowell in his study of Mars, and though he reported straight markings, they were few and shifting and were not at all like those described by Lowell. (Pickering had foibles of his own, however. Basing his beliefs on his detailed studies of the Moon, he was sure that it supported life, and this was, if anything, more startling than canals on Mars.)

Now, in the first decade of the twentieth century, the

two Bostonians entered a new field of rivalry, for each sought the trans-Neptunian planet. Lowell, doing his best to account for the anomalies in the orbits of Uranus and Neptune, engaged in enormous calculations and ended up with a trans-Neptunian planet that had an orbit that was quite tilted and quite elliptical. He estimated that its distance from the Sun varied from 5,100,000,000 kilometers (3,200,000,000 miles) at perihelion to 7,700,000,000 kilometers (4,400,000,000 miles) at aphelion. Pickering's orbit, obtained by less computation and more intuition, was distinctly farther from the Sun that Lowell's was.

Given the orbits, each man could predict the approximate position of his own theoretical trans-Neptunian planet at some particular time. In theory, one need only comb the sky in the indicated area and come up with the planet— but it wasn't that easy.

One could, in theory, look at every star in the region, record its position, and see if there was any record of that star on a star map. If there wasn't, it had wandered in from elsewhere and was a planet. That precise system had worked for Neptune, but for the new, much dimmer planet there were far too many stars to check. Even though twentieth-century astronomers had photography with which to record the positions of stars for study at leisure, which the Neptune-discoverers of 1846 didn't have, the method was not practical.

Lowell, working at Lowell Observatory, which he had built in the clear desert air of Flagstaff, Arizona, used another method. He took a photograph of a portion of the sky in the region where he thought the planet might be, then another photograph of the same region three days later. In three days, even the slow motion of a trans-Neptunian planet would have produced a noticeable shift in position.

Then, taking the pairs of pictures, he would compare the many stars of one to the many of the other in a slow and painstaking effort to see if one had changed its position. He did this over and over again for something like eleven years, bending over his plates endlessly, poring at them through a magnifying glass, studying the tiny dots, and comparing them.

Over and over he would find a shift. Each time his heart would leap, but each time the shift was too great and it turned out to be an asteroid. As more and more observations were made of Neptune and as its orbital discrepancies became better known, Lowell would recalculate his trans-Neptunian orbit and shift somewhat the focus of his most intense efforts. When he had to be away from the lab, his assistants carried on the search while he wrote to them constantly for news and went over all their plates when he returned, double-checking.*

He wore himself out, losing weight and equanimity, and died of a stroke on November 12, 1916, at the comparatively young age of sixty-one.

Pickering's efforts were not nearly as intense as Lowell's, but they reached a climax in the years following Lowell's death. At Mount Wilson Observatory in California, a young astronomer, Milton La Salle Humason, using Pickering's figures, began to search for the trans-Neptunian planet by the same system Lowell had used.

Humason did not succeed either, but he did not carry on too long. Lowell's long failure had taken the heart out of the search for most people and, after a while, Humason decided the planet wasn't there and the devil with it. In later years, looking back on the plates he had taken with the benefit of hindsight, it turned out he had photographed the trans-Neptunian planet twice. One time, a nearby star, brighter than the planet, had drowned it out. The second time, its image had just happened to fall on a tiny flaw in the plate.

Heartbreaking, but they don't pay off on heartbreak.

One person who did not give up was Percival Lowell. Dead he might be, but his money was not. He had left a trust fund for use in the search for the trans-Neptunian planet, and a decade after his death his brother Abbott Lawrence Lowell† added additional money to the fund.

By 1929 this money had made it possible to add a new telescope to the equipment at Lowell Observatory, one that had a very wide field and could photograph sharply all the

* There is no question in my mind that if he were asked what he would do if he knew he had only six months to live, he would have answered, "Look harder!"

† Abbott was president of Harvard University for a quarter of a century. Percival Lowell's sister was the poet Amy Lowell; his granduncle was the poet James Russell Lowell.

stars over a considerably larger area of the sky than had been possible before. Using an exposure of one hour, stars down to the seventeenth magnitude could be recorded and the trans-Neptunian planet, if it existed, was sure to be bright enough, and to spare, for detection.

Also added was a young astronomer named Clyde Willian Tombaugh. Tombaugh had been born in Streator, Illinois, on February 4, 1906, and his family was too poor to send him to college. He was sufficiently interested in astronomy, however, to build a telescope with a 9-inch lens, making use of parts of old machinery available on his father's farm. With his homemade telescope, he observed Mars carefully, saw the canals, and sent a report on his experience in building the telescope and on his observations therewith to Vesto Melvin Slipher, who was then director of Lowell Observatory. Tombaugh rightly felt the observatory would be interested in anything to do with the canals. Slipher was impressed enough to offer the young man a job and Tombaugh accepted it.

Tombaugh, still in his early twenties, young, vigorous, and full of enthusiasm, tackled the task that had so preoccupied Lowell and continued the search for the trans-Neptunian planet. He began to take photographs of the star field in Aries, Taurus, and Gemini, the area where Lowell's calculations had indicated the planet might be. Many thousands of stars were on each plate.

The task would have continued to be a virtually impossible one except for still another technical advance. Tombaugh had a "blink comparator" which Lowell had not had.

The blink comparator could project light through one plate taken on a certain day and then through the other plate taken a few days later and do so in rapid alternation. The plates were adjusted so that the stars on each were focused on the same spot. The true stars in the field would then all be precisely the same position relative to each other and would produce precisely the same projection. The rapid alternation would be so fast that the eye would not detect the flashing but would see a steady unchanging picture.

*If*, however, there was a planetary object present, it would have moved between the times the two photographs were taken and the effect of the blink comparator would be that of showing the planet in one position and then in a new position in rapid alternation. The planet would blink rapidly while all about it were motionless.

It was now not necessary to compare each one of the many thousands of stars on one plate with each one of the many thousands of stars on the other. It was only necessary to study every part of the plate in order to catch sight of that tiny blinking alternation and make sure the motion was too small for it to be the result of an asteroid.

Tombaugh began the search in the fall of 1929 and by February 1930 was working through the boundary region between Taurus and Gemini. Here the stars were particularly densely strewn and he found himself struggling with single plates that contained as many as 400,000 stars. He was having a miserable time and quite arbitrarily, just to give himself a rest, he switched to the other end of Gemini where the stars were sparser and his plates carried only 50,000 of them.

Then, at 4 P.M. of February 18, he spotted the blink. It was a fifteenth-magnitude object and the shift was a small one of 3.5 millimeters (⅛ inch). It could not be an asteroid. It had to be the trans-Neptunian planet. He looked for earlier photographs of the region to see if he could not spot a "star" that seemed to have been moving progressively. Knowing where to look, he had no trouble finding it.

From day to day he observed that object, and each day's motion proved more conclusively that it was what he had been looking for. The discovery of the ninth planet was formally announced on March 13, 1930, which was the one hundred forty-ninth anniversary of the discovery of Uranus and the seventy-fifth anniversary of the birth of Percival Lowell.

There were some suggestions that the new planet be named "Lowell," but that was not seriously considered. A mythological named was needed, and "Pluto" was adopted for the purpose.

It was an appropriate name, for the new planet, farther from the Sun than any other, was far enough out in the darkness of space to be named for the god of the dark underworld. In addition, the first two letters of the name were the initials of Percival Lowell, and don't think that those who proposed the name were unaware of that.

It is rather sad to think that if Percival Lowell had only lived to the thoroughly attainable age of seventy-five, he would have witnessed the discovery. Pickering, who died in 1938, just a month short of his eightieth birthday, did live to see it.

Yet, in some ways, the discovery was pure luck. The orbit of Pluto was markedly different from the orbit that Lowell had calculated. Pluto's orbit was considerably more titled and more elliptical than Lowell had supposed. What's more, Pluto's orbit was considerably closer to the Sun than Lowell had supposed. At aphelion it was 7,400,000,000 kilometers (4,600,000,000 miles) from the Sun, and at perihelion it was only 4,400,000,000 kilometers (2,700,000,-000 miles) from the Sun. At perihelion it was actually slightly closer to the sun than Neptune ever gets. (However, such is the tilt of Pluto's orbit that even when it seems to cross Neptune's orbit in the usual drawings of the solar system, it does so 1,400,000,000 kilometers (868,-000,000 miles) away in the third dimension. Where Lowell had expected the trans-Neptunian planet to circle the Sun once in 282 years (and where Pickering's figure was 373 years), Pluto's actual revolutionary period was 248 years.

It was luck that Pluto was in a part of its orbit which was comparatively close to Lowell's calculated position. If it had been in other parts of its orbit, it would have been so far from the point of calculation that the kind of search carried through by Lowell, Humason, and Tombaugh would not have succeeded.

The discrepancy in orbit could be dismissed, however, considering the uncertainty of the data with which Lowell had had to work. What was much more important was that Pluto was so *dim*. It was at least two magnitudes dimmer than it had been expected to be and it would not show a disc. Both facts could only be explained by supposing it to be considerably smaller than any of the other outer planets. It was not only far smaller than the larger giants Jupiter and Saturn, it was far smaller than the medium giants Uranus and Neptune.

In fact, the more closely it was studied, the smaller it seemed to be. For a while, it was thought to be as massive as the Earth but in recent years better data seemed to make it no more massive than Mars, or only one tenth as massive as Earth.

In early 1976, spectroscopic analyses of its light confirmed what had previously been supposed—that the planet was far enough from the Sun and therefore cold enough to have frozen methane coating its surface. But methane is frozen only at temperatures lower than 89° A. For a planetary surface temperature to remain that low, the planet must not only be far from the Sun, it must be small enough not to have developed much internal heat. Some astronomers now wonder if Pluto's mass might be no more than that of the Moon, or only one eightieth the mass of the Earth.

Whatever the actual mass, it is quite certain that Pluto is far too small to have captured any comets or to have any significant effect on the orbital movements of Uranus or Neptune. All the orbital discrepancies used to calculate the position of the trans-Neptunian planet have nothing to do with Pluto. Pluto's discovery is just an accidental fringe benefit of the search for the trans-Neptunian planet—like Columbus' discovery of America when he was heading for Asia.

But that means the trans-Neptunian planet (or trans-Plutonian planet, as it must now be called) that accounts for the orbital discrepancies must still exist and be out there somewhere. It is probably more distant than Pluto and must certainly be far more massive. Perhaps the size is great enough to make up for the greater distance so that it may not be much, if any, dimmer than Pluto and can be detected with no greater trouble—but I have the feeling no one is looking.

Well, we can wait. Neptune was discovered sixty-five years after Uranus was, and Pluto was discovered eighty-four years after Neptune was. If we allow a reasonable hundred-year gap for the trans-Plutonian planet, that takes us to 2030.

By then, assuming civilization survives, we ought to have a large telescope in orbit or on the Moon that can make observations without an interfering atmosphere. Furthermore, advanced computerization will probably allow the telescope to search for the blink without human interference and it would, in a matter of months, do what would have taken astronomers with the equipment of Tombaugh centuries, perhaps.

And then the trans-Plutonian planet can be found.

In fact, when the time comes that we can set up astronomical stations in the outer solar system, we may find several trans-Plutonian planets and the solar system will assume the vast size it must have in reality.

# VI

---

# OUR COSMOS

## Thirteen
# QUASAR, QUASAR, BURNING BRIGHT

Some months ago I received an unusual request. A charming young woman, who had met me at a convention and had been impressed by my suave demeanor, wrote to tell me that her twenty-fifth birthday was approaching. Her best friend, as it happened, would be celebrating her twenty-eighth birthday on the very same day.

Would it be possible, she asked, to celebrate the day by taking me out to lunch at the Russian Tea Room?

I hesitated. I live in a constant atmosphere of looming deadlines and have made long speeches to anyone who will listen on the iniquity of people who are forever expecting me to have lunch with them when I desperately need to remain glued to the typewriter. Yet somehow, having lunch with two young women in order to help them celebrate their birthdays is quite different from having a business lunch, right? And besides, the Russian Tea Room is one of my favorite restaurants, right?

So I finally agreed. All in good time, I got to the restaurant and found the two young women waiting for me. They clapped their little hands in glee and I sat down between them, feeling pretty gleeful myself. We had a very pleasant time indeed, talking, joking, laughing, and when dessert time came, I prepared to order my inevitable baklava.

The establishment, however, had somehow got the notion that a birthday was involved and they beat me to it. Two waiters came out with a cake bearing a candle. They sang "Happy Birthday to You" and placed the cake right in front of me.

I could see why they did so. If you yourself were to see a man in his late youth flanked on either side by an attractive young woman and knew that there was a birthday involved, wouldn't you suppose it was the man who was getting a special birthday treat?

I don't like inaccuracy, however, so I smiled genially at the waiters and said, "No, no. It is the young women who are having the birthday. *I* am the birthday present."

The look of awed respect in the eyes of the waiters was beautiful to see. —But you know me. I just sat there trying to look modest.

The moral is that things are not always what they seem—which brings me to the subject of the article.

The first astronomer who tried to map the heavens and indicate the position of at least some of the various stars was Hipparchus of Nicaea. He prepared a map about 130 B.C. on which he listed 1,080 stars, giving the celestial latitude and longitude of each, as best as they could be determined without either a telescope or a modern clock.

The position of a star was one of the two stellar properties that could be determined without modern instruments. The other was the relative brightness and some stars, after all, are brighter than others. Hipparchus didn't neglect that fact.

He divided the stars into six classes. The first class included the twenty brightest stars in the sky. The second included stars dimmer than those; and the third, stars that were still dimmer. Then there followed the fourth, fifth, and sixth classes, the last of which included those stars just barely visible on a dark, moonless night to a person with sharp vision.

Each class eventually came to be called a "magnitude," from the Latin word for "great." It was only natural to use that word, since throughout ancient and medieval times the stars were supposed to be all at the same distance —all stuck to the hard material of the "firmament" like luminous thumbtacks. It was almost as if they were tiny holes in the firmament through which the glorious light of heaven could be seen, and the difference in brightness would then depend on the size or greatness of the hole.

The brightest stars, therefore, were of the "first magnitude," the next brightest were of the "second magnitude," and so on.

Hipparchus' works did not survive into modern times, but nearly three centuries after his time, another astronomer, Claudius Ptolemaeus, or Ptolemy, of Alexandria published a survey of the astronomical knowledge of the day, based largely on Hipparchus' work. Ptolemy included

Hipparchus' map, with some corrections, together with the notion of magnitudes. Since Ptolemy's work survived down to the present, we still retain the division of stars into magnitudes today.

The division of the stars into magnitudes was purely qualitative at first. Some first-magnitude stars are clearly brighter than other first-magnitude stars, but no account was taken of that. Nor did astronomers worry overmuch that the dimmest first-magnitude stars were not very much brighter than the brightest second-magnitude stars. There is, in fact, a continuous gradation of brightness among the stars, but the classification into discrete classes obscures that.

In the 1830s attempts began to be made to improve on the Hipparchus/Ptolemy system, which was, by then, two thousand years old.

One pioneer was the English astronomer John Herschel, who was observing the southern stars from the Cape of Good Hope. In 1836 he devised an instrument which would produce a small image of the full Moon that could be brightened or dimmed by the manipulation of a lens. The image could then be made equal in brightness to the image of a particular star. In this way, Herschel could estimate the comparative brightness of stars quite finely and could determine gradations smaller than a whole magnitude.

Using the full Moon, however, restricted the times when measurements could be made, and only the brighter stars could be measured since the dimmer ones were washed out in the moonlight.

At about the same time, however, a German physicist, Carl August von Steinheil, had worked out a similar device that could bring into juxtaposition the images of two different stars, one of which could be dimmed or brightened to match the other. This was the real birth of "stellar photometry" and, for the first time, magnitudes could be measured by objective instrument rather than by subjective estimation by eye alone.

Once this came to pass, it became important to determine the significance of magnitude. How does brightness change as one goes up or down the scale of magnitude?

To the eye, it seems that the change in brightness from each magnitude to the next is the same. One goes from first magnitude to sixth magnitude in equal steps.

But can these steps be represented as though we went up the number scale 1, 2, 3, 4, 5, 6? Was the sixth magnitude 1, the fifth magnitude 2, the fourth magnitude 3 and so on? Was a difference of one magnitude equivalent to a doubling of brightness, a difference of two magnitudes to a tripling, a difference of three magnitudes to a quadrupling, and so on? If that were so, brightness would increase up the steps of the magnitudes in equal increments and we would have an "arithmetic progression."

Steinheil did not think this was so. He thought the progression was by equal ratios. In other words, if the sixth-magnitude star was 1 and the fifth-magnitude was 2, then the fourth-magnitude star would be 4, the third-magnitude star 8, the second-magnitude star 16, and the first-magnitude star 32. This is a "logarithmic progression."

Steinheil was right and, as time went on, physiologists showed that the human senses, generally, work logarithmically. You can see this for yourself if you have a three-way light bulb which has a 50-, 100-, and 150-watt set of levels. Switch from the 50- to the 100-watt level and there is a marked brightening. Pass on to the 150-watt level and the further brightening seems considerably less even though there has been another 50-watt increase. Your visual sense detects a 100 per cent increase in the first step and only a 50 per cent increase in the second.

Similarly, you can easily tell the difference between a 1-pound weight and a 2-pound weight of the same dimensions by hefting them. You cannot easily distinguish between a 30-pound weight and a 31-pound weight in the same way, even though the difference is still 1 pound. In the first case you are detecting a 100 per cent difference; in the second you are failing to detect a 3 per cent difference.

Of course, it would be too much to expect that a system of magnitudes, chosen by eye by Hipparchus, would happen to divide stars into groups each of which was just twice the brightness of the next group below. The ratio would be some less convenient value, surely.

In 1856, the English astronomer Norman Robert Pogson pointed out that the average first-magnitude star is about a hundred times as bright as the average sixth-magnitude star, judging by photometry. In order to make the five intervals between the magnitudes come out to just 100, we must make the ratio of each of the five intervals the fifth root of 100, which comes out to about 2.512. (In other words 2.512 × 2.512 × 2.512 × 2.512 × 2.512 × 2.512 is just about equal to 100.)

Therefore, if you choose a magnitude of 1.0 in such a way that some of the traditional first-magnitude stars are brighter than that, and some dimmer, you can then proceed to work your way down by ratios of 2.512.

As photometers improved, astronomers could determine magnitudes to one decimal and even, on occasion, could make a stab at the second decimal. The brighter of two stars separated by a tenth of a magnitude is about 1.1 times as bright than the dimmer one. The brighter of two stars separated by a hundredth of a magnitude is about 1.01 times as bright as the dimmer.

Using the new system, we no longer have to say that Pollux and Fomalhaut are both first-magnitude stars. Instead, we can say that Pollux has a magnitude of 1.16 and Fomalhaut a magnitude of 1.19. This means that Pollux, with the lower number, is brighter than Fomalhaut by 0.03 magnitudes.

We might say that we can call any star with a magnitude between 1.5 and 2.5 a second-magnitude star. Working our way downward from that, any star with a magnitude between 2.5 and 3.5 would be a third-magnitude star, and so on. Stars with a magnitude between 5.5 and 6.5 would be sixth-magnitude stars and would belong to the class originally defined as the dimmest stars that could be seen.

By the time Pogson had worked out his magnitude scale, however, the sixth-magnitude stars were by no means the dimmest that could be seen. The telescope revealed far dimmer stars and successive improvements of the instrument revealed still dimmer ones.

That didn't matter, however. By continuing to use that ratio of 2.512, we can have seventh-magnitude stars, eighth-magnitude stars, ninth-magnitude stars, and so on, measuring each to as close a value as our instruments will allow us to.

The best contemporary telescopes will reveal stars as dim as the twentieth magnitude if we place our eye to the eyepiece. If we place a photographic plate there instead and let the focused light accumulate, we can detect stars down to the twenty-fourth magnitude.

That's not bad, really, since a twenty-fourth-magnitude object is eighteen magnitudes dimmer than the dimmest object we can see with the unaided eye. By our logarithmic scale, this means that the dimmest star the ancients could see is about 16,000,000 times as bright as the dimmest star *we* can see.

We started from the second magnitude a few paragraphs back and worked our way dimward. Let's start from there again and work our way brightward. If stars with magnitudes from 1.5 to 2.5 are second-magnitude, then stars with magnitudes from 0.5 to 1.5 are first magnitude.

But there are no less than eight stars with magnitudes lower than 0.5. What are they as far as magnitude is concerned? Some stars are even brighter than would be represented by a magnitude of 0.0 and must have their magnitudes expressed as negative numbers. Can we speak of the "zeroth magnitude" and define it as lying between magnitudes from −0.5 to 0.5? There are six stars of the zeroth magnitude, ranging from Procyon with a magnitude of 0.38 to Alpha Centauri with a magnitude of −0.27.

There are even two stars with magnitudes beyond −0.5 and which, therefore, are of the "minus-first magnitude." They are Canopus with a magnitude of −0.72 and Sirius with a magnitude of −1.42.

Astronomers, however, can't break with tradition quite that far. They can bring themselves to go beyond Hipparchus' sixth magnitude but not beyond his first magnitude. The stars with magnitudes of less than 0.5, even Sirius, are all lumped together among the first-magnitude stars.

It means that the brightest traditional first-magnitude star, Sirius, is actually three magnitudes brighter than the dimmest traditional first-magnitude star, Castor, whose magnitude of 1.58 actually puts it just over the edge into the second magnitude. In terms of brightness, Sirius is about sixteen times as bright as Castor and is about 15,000,000,000 times as bright as the faintest star our telescopes can show us.

Are there objects in the sky that are brighter than Sirius?

Certainly! Hipparchus restricted his magnitude classification to stars, but now that magnitudes have been reduced to numbers and ratios, astronomers can continue to move along the scale of negative numbers and up the level of brightness as far as they want to.

Thus, when the planet Jupiter is at its brightest, it reaches a magnitude of −2.5. No astronomer speaks of it as being the "minus-second magnitude" or of any named magnitude—but the number can be given. Then, Mars can reach a magnitude of −2.8, while Venus, the brightest jewel in the sky, can attain a magnitude of −4.3. At its brightest, Venus is about fifteen times as bright as Sirius.

And even that doesn't represent the top. The Moon is far brighter even that Venus and at full Moon attains a total magnitude of −12.6. That means the full Moon is about two thousand times as bright as Venus.

This leaves the Sun, whose magnitude is −26.91. The Sun is thus 525,000 times as bright as the full Moon, 1,000,000,000 times as bright as Venus, 15,000,000,000 times as bright as Sirius, and 250,000,000,000,000,000,000 times as bright as the dimmest object the telescope will show us.

And since there is nothing brighter to see in the sky than the Sun and nothing dimmer than the dimmest star present-day telescopes can show us, we have reached the limit in both directions, having traversed a range of fifty-one magnitudes.

But, as I said in the introduction to this article, things are not always what they seem.

All these magnitudes I've been talking about are *apparent* magnitudes. The brightness of an object depends not only on how much light it emits, but also on how distant it is from us. An object that is actually extraordinarily dim in an absolute sense, like a 100-watt light bulb, can be placed just behind our shoulder and it can then be brighter to our eyes than the Moon is. On the other hand, a star that gives out much more light than the Sun can be so far away that not even a telescope will show it to us.

In order, then, to determine levels of *real* brightness, to measure the light an object *actually* emits—its "luminosity"—we must imagine that all the objects in question are at some fixed distance from us. The fixed distance has been selected (arbitrarily) as 10 parsecs (32.6 light-years).

Once the distance of any luminous object is known and its brightness at that distance is measured, we can calculate what its brightness would be at any other distance. The magnitude an object would have at 10 parsecs is its "absolute magnitude."

Our Sun, for instance, is about 150,000,000 kilometers (93,000,000 miles) from us, or 1/200,000 of a parsec. Imagine it out to 10 parsecs and you have increased its distance by 2,000,000 times. Its apparent brightness sinks by the square of that number, or 4,000,000,000,000 times. That means its brightness sinks by about thirty-one and a half magnitudes. Its absolute magnitude is thus about 4.7. The Sun, seen from a distance of 10 parsecs would be visible, but it would shine as a fairly dim and quite unremarkable star.

What about Sirius? It is already at a distance of 2.65 parsecs. If we imagined it out at 10 parsecs, its brightness would dim by nearly three magnitudes, and its absolute magnitude would be 1.3. It would no longer be the brightest star in the sky, but it would still be a first-magnitude star.

The absolute magnitudes, which wipe out difference in distance as a factor, show us that Sirius is about twenty-three times as luminous as the Sun, that is, it emits twenty-three times as much light.

Sirius is far from the most luminous star there is, however. There are stars that do much better. Of all the first-magnitude stars, the most distant is Rigel, which is 165 parsecs away. It is only the seventh brightest star in the sky and is only one quarter as bright as Sirius. Still, Rigel is over sixty times as far from us as Sirius is. To make so respectable a show from such a distance, Rigel must be very luminous.

And indeed it is. The absolute magnitude of Rigel is −6.2. Place it at a distance of 10 parsecs, and even though it would be at nearly four times the actual distance of Sirius, it would not only greatly outshine that star, it would shine even brighter than Venus—about six times as bright. In fact, Rigel is 1,000 times as luminous as Sirius and 23,000 times as luminous as the Sun.

Even Rigel isn't the record-holder. It is the most luminous star that we know of in our own Galaxy, but there are other galaxies. The Larger Magellanic Cloud is a kind of satellite-galaxy of our own and in it is a star called "S Doradus." It is too dim to see except with a telescope, but it is some 45,000 parsecs away and astronomers were astonished it was as bright as it was, considering its distance. It turns out to have an absolute magnitude of −9.5. That makes it about twenty-one times as luminous as Rigel and nearly half a millions times as luminous as our Sun.

If S Doradus were in place of our Sun, a planet circling it at seventeen times the distance of Pluto would see it shine as brightly as we see our Sun shine.

S Doradus is the most luminous, stable star that we know of; it emits more light day after day, century after century, than any other. Not all stars are stable, however. Occasionally, stars explode into "novas" and then gain luminosity sharply, if temporarily.

The size of the gain depends on the size of the star. The more massive the star, the more enormous the explosion. The really magnificent explosion of a "supernova" can bring a single massive star to an absolute magnitude, very briefly, of about −19.

For a brief time, such a supernova will be shining with a luminosity some 6,000 times as great as S Doradus and about 10,000,000,000 times that of our Sun. Even at a distance of 10 parsecs it will shine 360 times as brightly as the full Moon, though it would be only a thousandth as bright as the Sun.

Do we now have a record for luminosity?

Perhaps not. A supernova is only a single star. Might we not consider the luminosity of a group of stars?

A pair of stars, reasonably close together, looks like a single star from a distance. If both stars are of equal brightness, the combination is 0.75 magnitudes brighter than either star singly.

Double stars are very common, and even triple and quadruple star systems are not exactly rare. In fact, stars exist in large clusters as well. There are about 125 known "globular clusters" associated with our Galaxy, and each contains anywhere from ten thousand to several hundred thousand stars, all densely packed together by the standards of our stellar neighborhood.

Well, then, suppose we consider a globular cluster made up of a million stars, each with the luminosity of our Sun. We could calculate its absolute magnitude to me −10.3. Such an enormous cluster would be, nevertheless, merely twice as luminous as the single star S Doradus. A gigantic supernova can attain a luminosity equal to 3,000 times that of a large globular cluster. No globular cluster can, therefore, set a luminosity record.

A galaxy itself, however, has at its nucleus, the equivalent of a globular cluster of enormous size. The center of our own Galaxy is a densely packed globular cluster made up of 100,000,000,000 stars. Its absolute magnitude can be calculated to be −22.8. (The rest of the Galaxy, outside the nucleus, has a relatively sparse scattering of stars, and if its luminosity is included, the over-all value may reach −22.9.)

*That* looks like a new record. The galactic nucleus shines with a luminosity over three times that of a supernova at a peak. (Still, this is not a very great difference in luminosity, and when a gigantic supernova blazes out in a particular galaxy, it is quite likely to give out as much light, at its peak, as all the rest of the galaxy combined.)

Of course, our Galaxy is not the largest there is. A large galaxy can easily be ten times the mass of our own and might have an absolute magnitude of −25.

There is a catch to this calculation of the absolute magnitudes of globular clusters and galaxies, however, for we are dealing with extended bodies. A large globular cluster could be up to 100 parsecs across, and a galactic nucleus can be up to 5,000 parsecs across. The absolute magnitude can be calculated, but it can't be experienced in the ordinary way.

If you imagined the central point of a globular cluster or of a galactic nucleus to be 10 parsecs away, you would be *within* the object. You would see stars all about you and you would not have the sense of a combined luminosity, anymore than you have it now in our own Galaxy.

To be sure, we might use 1,000,000 parsecs as the conventional distance in measuring luminosity, and then we could see that a large galaxy outshines any individual star under any circumstance. But then, all the objects seen at that distance would seem very dim and unimpressive.

If we want to look for a record beyond a supernova, then we must ask if there is anything that would look like a single object of reasonably small size at a distance of 10 parsecs and which would yet outshine a supernova from day to day, steadily.

There *is* an answer. What we call "quasars" are, apparently, galactic nuclei so condensed and so brilliant that they can be seen (telescopically) at distance of hundreds of millions of parsecs. No other object can be seen at such distances. A typical quasar is thought to be, perhaps, only half a parsec or so in diameter, and yet it shines with the luminosity of a hundred galaxies such as our own.

Half a parsec is a respectable diameter; it is about 12,000,000 times the diameter of our Sun; it is over 1,000 times the diameter of the orbit of Pluto. Place a quasar at a distance of 10 parsecs and its apparent diameter will be nearly 3 degrees across. That is about six times the diameter of our Sun or full Moon, but we would still see it as a single blazing object.

The average quasar will then have an absolute magnitude of −28. It will shine, *even at the distance of 10 parsecs* about twice as brilliantly as the Sun does in our sky, even though the quasar is 2,000,000 times as far away.

The question then arises as to what the brightest quasar might be. All quasars tend to vary in brightness quite a bit from time to time. They appear in our telescopes as ordinary-seeming, very dim stars (thanks to their great distance) and were photographed in past years long before they were known to be special. (The discovery was made through their intense radio-wave emission.) If astronomers go back through past records, astonishing peaks of luminosity may show up.

In 1975 two Harvard astronomers, Lola J. Eachus and William Liller, traced back quasar 3C279. It usually shines with an apparent magnitude of 18, but back in 1937 it briefly attained an apparent magnitude of 11.

To shine as brightly as the eleventh magnitude from a distance of about 2,000,000,000 parsecs is incredible, almost. At its peak, 3C279 shone with the light of ten thousand ordinary galaxies and its absolute magnitude was calculated by Eachus and Liller to have reached a peak of −31.

Imagine 3C279 at a distance of 10 parsecs from us and it would shine with a brilliance of forty times that of our Sun as we see it now.

A quasar such as 3C279 can reach a peak luminosity, then, of 100,000,000,000,000,000 times that of our Sun—or 500,000,000 times that of S Doradus—or over 60,000 times that of a vast supernova at its peak—or 1,000 times that of our entire Galaxy taken as a unit.

And that's the record, as far as we know now.

## Fourteen
# THE DARK COMPANION

Opportunities for embarrassment come my way as they come anyone else's, and I don't always avoid them.

Although I am well known as a hard-nosed scientist, unapt to accept rubbish (see Chapter 17), I welcome new ideas if they are advanced by people who know what they are talking about and who respect rationality—and there we have the potentiality for embarrassment.

First, a couple of cases where I was *not* embarrassed—

In 1974 a book called *The Jupiter Effect*, by John Gribbin and Stephen Plagemann, was published by Walker & Co. It dealt with the possible effect of planetary position on solar tides, thence on the solar wind, thence on plate tectonics on Earth, thence on earthquakes in California. It was tenuous reasoning that arrived at a very iffy conclusion, but it seemed to me to be the work of honest and logical men, so when I was asked to write an introduction to the book, I did so. My name was eventually placed on the book jacket as prominently as the authors'.

The book was roughly handled by many reviewers, as I expected it might be, and my good friend Lester del Rey never tires of calling me "an astrologer" because of the introduction, but I stand by my guns. The book was worth a hearing and I'm not ashamed of the association.

Then, in 1976, Doubleday published *The Fire Came By*, by John Baxter and Thomas Atkins. It was a study in depth of the great Siberian explosion of 1908, which had long been assumed to have been caused by a meteorite. The authors take into account all the evidence they could gather and discussed all the explanations that have been offered once it became apparent that there was no sign of any crater or meteoric fragments. They end their study by suggesting that the explosion was caused by an extra-terrestrial, nuclear-powered vessel that went out of control and crashed into the Earth.

Larry Ashmead, then of Doubleday, asked me to look at the manuscript for a possible favorable comment but did so with considerable reluctance, for he assumed I would tear it up once I had read it.

I didn't. I found the book fascinating and honest and, in my opinion, worth a hearing.* I *asked* Doubleday if I might do an introduction, and when they said I might, I wrote one. My name appears on the book jacket almost as prominently as those of the authors, and though I dare say that few astronomers will take the book seriously, I will again stick to my guns. I am not ashamed of this association, either.

And now for the embarrassment—

A book called *The Sirius Mystery* has just been published.† It deals with a West African tribe whose traditions seem to include knowledge of the satellites of Jupiter, of the rings of Saturn, and of the white-dwarf companion of Sirius—knowledge they seem to attribute to travelers from a planet circling Sirius.

While the book was still in manuscript form, the author got in touch with me, described the thesis of the book, and asked me to read it so that I might give it some favorable comment. Very reluctantly I agreed to let him send me the manuscript. —After all, I can't very well refuse to *look* at what a man has to say.

The manuscript arrived, and I tried to read it. I hate to be unpleasant and insulting, for the author in his contact with me seemed a pleasant and sincere man—but the plain fact is that I found the book unreadable, and what I managed to choke down I found unconvincing.

Therefore, I refused any comment.

The author called me at one time thereafter and rather put pressure on me to reconsider. I find it difficult to be rude, but I managed to maintain my refusal in an uneasily polite manner.

Then, he said, "Well, did you detect any errors?"

Of course, I hadn't. I had read only a small portion of the book, a portion in which he talked about this West African tribe concerning which I knew nothing. He could have

* My friend James Oberg, who has studied the matter closely, thinks I am overgenerous in this opinion, and he may be right.

† Since I have nothing good to say of it, I will not name the author or publisher.

said anything at all without my being able to point out a definite *error*. So, to get rid of him and to be polite, I said, "No, I did not detect any errors."

I got what I deserved. That's what I said and I did not specify that I was not to be quoted—so when the book was published and advertisements appeared in the newspapers, there I was, quoted as telling the world there were no errors in the book.

I am embarrassed at my stupidity, but I assure you I will never be caught that way again.

I will console myself, somewhat, by telling you the tale of the discovery of the white-dwarf companion of Sirius by modern astronomers, who did it without the help of extraterrestrial visitors.‡ It is a drama in three acts.

### *Act I—Friedrich Wilhelm Bessel—1844*

Friedrich Wilhelm Bessel was born in Minden, Prussia, on July 22, 1784. He made his living as an accountant at first, but taught himself astronomy and at the age of twenty recalculated the orbit of Halley's comet with such elegance that the German astronomer Heinrich W. M. Olbers was sufficiently impressed to get him a job at an observatory.

In the 1830s Bessel was engaged in the great astronomic adventure of that decade, the attempt to determine the distance from the Sun of some star. To accomplish the task, astronomers had to choose a star that was comparatively near Earth and note the steady, elliptical shift in its position (parallax) compared to farther ones, as Earth revolved in its orbit. But how could one pick out a nearby star when one didn't know beforehand which were near and which were far?

One must guess and there were two possible clues. First, a bright star was more likely to be close than a dim star was, since proximity could be the cause of the brightness. Second, a star which shifted position (proper motion) considerably from year to year relative to the other stars was

‡ I went into the subject briefly in "Twinkle, Twinkle, Little Star," in *Adding a Dimension* (Doubleday, 1964), but that was a seventh of a century ago and I'm passing through in more detail this time, and in a different direction.

likely to be closer than one which shifted little or not at all, since proximity tended to magnify the extent of the shift.

Two other investigators, the Scottish astronomer Thomas Henderson and the German-Russian Friedrich G. W. von Struve, went for brightness. Henderson, at Capetown, South Africa, took aim at Alpha Centauri, the brightest of the southern stars. Von Struve zeroed in on Vega, the brightest of the far northern stars.

Bessel went for fast motion and chose 61 Cygni, a rather dim star (fifth magnitude) but one which happened to have the fastest proper motion known at the time.

All three men were successful, but Bessel announced his results first, in 1838, and he gets credit for being the first to determine the distance of a star.

Having won this victory, Bessel was ready to determine other distances, and eventually he decided to tackle Sirius. Sirius is the brightest of all the stars and therefore could well be one of our neighbors. (It is. Its distance is only two fifths that of 61 Cygni.) It also has a fairly rapid proper motion.

Determining the parallax of a star isn't easy, however. If a star were absolutely motionless, and if the Earth's motion were absolutely regular, and if the speed of light were infinite and there were no atmosphere, it might be—but all that is not the way it is, so it isn't.

A star might have a parallax and mark out an ellipse, but it also has a proper motion in a straight line. The combination of the straight proper motion and the elliptical parallax produces a wavy motion, which is further complicated by atmospheric refraction, by light abberation, by various wobbles in Earth's motion, and so on. Every possible interference has to be allowed for and subtracted and, finally, when everything has been subtracted, what's left over is parallax. Of course, since every subtraction has its errors, the parallax with which you're left can be pretty fuzzy.

Bessel got to work on Sirius, observing it from night to night, checking the other reported observations, and so on. He allowed for all the minor factors, subtracted the proper motion, and got an ellipse—but it was not a parallax ellipse.

The ellipse marked out by a star's parallax has to be one

that completes its turn in one year since a parallax reflects the yearly turn of Earth about the Sun. This did not happen in Sirius' case.

Bessel could easily tell that the ellipse he ended with took far more than a year to complete. In fact, at the rate Sirius was marking out that ellipse, it was going to take fifty years to complete it. So it wasn't parallax. It was something else.

There *was* something else that could make a star move in a long-turning ellipse like that. The star might be a binary; it might be one of a pair that turned about each other, pivoting on the center of gravity of the system. William Herschel (see Chapter 10) had discovered such binaries in 1784, and they were by no means uncommon.

Why, then, shouldn't Sirius be a member of a binary system? It was just moving about a center of gravity, with another star always on the opposite side.

That was great, but there was a catch. Bessel couldn't see the other star. He knew exactly where it ought to be at all times, from Sirius' movements and the law of gravity, but it just wasn't there.

Was the other star actually a planet? One could not possibly see planets at the distance of Sirius. (From Sirius one could see our Sun, but not Jupiter.)

It couldn't be a planet. The only reason you can't see planets is that they're too small to shine like a star; and if they're too small to do that, they're also too small to dispose of enough of a gravitational field to throw Sirius about like that. The other member of the binary *had* to be a star. It just couldn't be seen, even though it *was* a star.

In Bessel's time that was not such an unbelievable thing. Notions of the law of conservation of energy were in the air at that time and it seemed reasonable to assume that a star only had a finite quantity of energy at its disposal. If so, a star could burn out just as a candle could. It took longer for a star to do so, but the principle was the same. Well, then, Sirius was accompanied by a star that had burned out, so of course it could not be seen.

In 1844 Bessel announced his discovery that Sirius had a dark Companion. (Later on, he found that the bright star Procyon also had a dark companion.)

Bessel died in Königsberg, Prussia, on March 17, 1846, and did not live to witness Act II of the drama.

Alvan Graham Clark was born in Fall River, Massachusetts, on July 10, 1832. His father, Alvan Clark, was a portrait painter who was fascinated by astronomy and lusted to grind lenses. (I don't understand the ecstasy of lens grinding myself, but the history of astronomy is littered with peculiar people who would rather grind lenses than eat.)

In the early nineteenth century, however, all lens grinders were British, French, or German, and no self-respecting European astronomer could even conceive of an American turning out anything useful in that respect. The elder Clark let his work speak for itself. He ground lenses, placed them in telescopes that he used himself to make excellent observations, which he reported. European astronomers, curious as to the instruments Clark used, learned he had ground the lenses himself and began to place orders with him.

By 1859 Clark was a celebrity. He was invited to London, where the greatest British astronomers were delighted to meet him. He returned to the United States and established a telescope factory in Cambridge, Massachusetts. His younger son, Alvan Graham Clark, worked with him.

In 1860 the Chancellor of the University of Mississippi wanted a good telescope with which to put the institution on the astronomical map. Since he was a Massachusetts man, he thought of the Clarks and placed the order with them. The Clarks got to work at once. (Alas, the telescope never got to Mississippi. Within a year, the Civil War was on and Mississippi was enemy territory. The telescope, when completed, went to the University of Chicago, instead.)

By 1862 Alvan Graham Clark had a lens that was ground to a fare-thee-well and looked beautiful. The next step was to check it in practice.

Clark placed the lens in a telescope, pointed it at Sirius, and took a good look at it. If the lens were perfect, he would see Sirius as a hard, bright, sharp point (allowing for a twinkle, if seeing wasn't very good). On the other hand, a tiny irregularity in the lens would blur or distort the point.

Clark looked and was chagrined to find a tiny spark of

light in the vicinity of Sirius where no spark of light should be. The easy conclusion was that there was an irregularity to the lens that split off a tiny speck of Sirius' light.

And yet when Clark looked anywhere else in the sky, there were no apparent problems, and nothing he could do to the lens in the way of further perfecting its shape could make that spark of light near Sirius disappear. He finally decided he saw the spark of light because it existed. There was something there. The spark of light was in the position where the dark Companion of Sirius should exist and that was it. He was seeing the Companion.

The Companion was not really very dim, for it had a magnitude of 7.1, almost bright enough for naked-eye visibility. It was, however, very close to Sirius, which was about 6,000 times brighter and which blanked it out. It took a good lens to make out that dim spark in the face of that nearby brilliance, so the trouble with Clark's lens was not its imperfection but its excellence.

One could no longer talk about the dark Companion of Sirius. It was now a dim Companion. That didn't upset the apple cart too much, however. If the Companion was not exactly a dead cinder, it was apparently a dying one on its last flicker.

Alvan Graham Clark died in Cambridge on June 9, 1897, and he did not live to witness Act III in the drama.

### Act III—Walter Sydney Adams—1915, 1922

Walter Sydney Adams was the son of an American missionary couple working in the Middle East. He was born in Antioch, Syria (then part of Turkey), on December 20, 1876, and was not brought to the United States till he was nine years old. After graduating from Dartmouth College in 1898 and following that with postgraduate training in Germany, he became an astronomer.

By this time, astronomy had been revolutionized by the use of the spectroscope. Astronomers were no longer restricted to noting the brightness and over-all color of a star's light. That light could be spread out into a spectrum crossed by dark lines. From the positions and patterns of the dark lines, the chemical elements present in the star could be determined. From slight shifts in position as compared with the lines produced by the same element in the laboratory, one could determine whether the star was approaching or receding, and how quickly.

In 1893 the German physicist Wilhelm Wien had shown how spectra varied with temperature. It was now possible to study the spectrum produced by a star and determine its surface temperature. For instance, our Sun has a surface temperature of 6,000° C., but Sirius is a much hotter star and has a surface temperature of 11,000° C.

It was clear that stars differed in color because the pattern of wavelengths they emitted varied with temperature. No matter what the structure or composition of a star, if its surface temperature was 2,500° C., it was red; if it was 4,500° C., it was orange; if it was 6,000° C., it was yellow-white; if it was 11,000° C., it was pure white; if it was 25,000° C., it was blue-white.

To Adams this raised an interesting problem. The Companion of Sirius had now been known for seventy years and had always been viewed as a dead or dying star. But if the Companion were dying and flickering toward extinction, it ought to be cool and therefore red in color. The trouble was that it was not red in color at all; it was white. It had to be hot, therefore, and it was hard to see how, in that case, it could be flickering out.

To reach a decision safely, what one needed was the spectrum of the Companion. Getting the spectrum of a seventh-magnitude star in the very teeth of a neighboring star with a magnitude of −1.42 was no mean trick. In 1915, however, Adams managed to do it.

The spectrum removed all doubt. The Companion was almost as hot as Sirius itself was. It had a surface temperature of about 10,000° C. and so it was considerably hotter even than our own Sun.

But that raised another question. If the Companion was almost as hot as Sirius was, then any given portion of the surface of the Companion should be nearly as brilliant as an equivalent portion of Sirius. Why, then, was it that the Companion was only 1/6,000 as bright as Sirius altogether?

The only reasonable answer to that question was that though, portion for portion, the surface of the Companion was nearly as bright as the surface of Sirius, there was far less, *far* less, total surface in the Companion.

In fact, if we know how bright a portion of the surface

of a star ought to be from its temperature, it is possible to calculate the surface area of the star that would account for that apparent brightness, and from that, in turn, we can calculate the star's diameter. It turns out, for instance, that the diameter of Sirius is 2,500,000 kilometers (1,500,-000 miles), or 1.8 times that of our Sun, while the diameter of the Companion is 47,000 kilometers (29,000 miles), or 0.033 that of our Sun.

The diameter of the Companion came as a shock, since the thought of a star that small seemed ludicrous. It was not just that it was smaller than our Sun; it was considerably smaller, even, than the planet Jupiter. It was, in fact, approximately the size of the planet Uranus.

Since the Companion was both white in color and dwarfish in size, it was called a "white dwarf" and was the first of a new class of stars, one that turned out to be fairly common. The Companion of Procyon, for instance, turned out to be a white dwarf, also.

As it happens, Sirius is sometimes called the "Dog Star," since it is the brightest star of the constellation of Canis Major, the "Great Dog." In view of that, some people took to calling the dwarfish companion "The Pup," a piece of cutesiness that is beneath contempt. The proper practice these days is to assign the stars of a multiple system letters of the alphabet in the order of brightness. Sirius is now called Sirius A and the Companion is Sirius B. (For this article I'll stick to Sirius and the Companion, however.)

The small size of the Companion is peculiar in itself, but what made it even more striking is that in other respects, the Companion is full-sized. From the distance between Sirius and the Companion and from the orbital period, it is possible to calculate that the total mass of the two stars is about 3.5 times that of the Sun. From the distance of each from the center of gravity, it can be shown that Sirius is 2.5 times the mass of the Sun and the Companion just about equal to the mass of the Sun.

But if the Companion has the mass of the Sun crammed into a globe with only one thirtieth the diameter (and therefore 1/9,000 the volume) of the Sun, then the average density of matter in the Companion must be 9,000 times that of the Sun, or 12,600 grams per cubic centimeter, or about 575 times the density of platinum.

If Adams had announced his findings no more than five years before, they would have been thrown out of court. Such a density figure would have seemed so ridiculous that the whole system of measuring temperature by spectroscopy might have been questioned and, perhaps, discarded.

In 1911, however, the New Zealand-born, Cambridge University physicist Ernest Rutherford had announced his theory of the nuclear atom, based on his observations of the behavior of atoms under bombardment by the newly discovered subatomic radiations of radioactive elements. It became clear that atoms were mostly empty space and that almost the entire mass of each atom was concentrated in a tiny nucleus taking up only about 1/1,000,000,000,-000,000 of the total space of the atom. It was easy to suppose the Companion and all white dwarfs to be made up of shattered atoms, so that the massive nuclei drew closer together than they possibly could when part of intact atoms.

In those circumstances, the density of the Companion made sense. Indeed, far greater densities could be possible.

Next came something else. The nature of the Companion having been worked out, it became possible to use it to prove something else even more esoteric.

In 1916 Albert Einstein had worked out his Theory of General Relativity. It made necessary the existence of three interesting phenomena for which the older Newtonian theory of gravitation had no room. Only one had already been observed—the anomalous advance of the perihelion of Mercury (see "The Planet That Wasn't," in the book of that name, Doubleday, 1976). But what of the other two?

The second was that light would bend in its path when moving through a gravitational field. This bending was very slight but might be just detectable using a gravitational field as intense as the Sun's.

As it happened, on May 29, 1919, a solar eclipse was scheduled to take place at just the time when more bright stars were in the vicinity of the eclipsed Sun than would be there at any other time of year. The Royal Astronomical Society of London made ready an expedition to the island of Principe, one that was designed to test Einstein's theory.

The positions of the stars near the Sun were carefully measured during the eclipse. If the Einstein light-bending took place, each star would appear to be just a trifle farther from the Sun than it should be. The extent of the shift away would depend on the apparent distance from the Sun. It was tedious and difficult work and the results were not altogether clear-cut, but on the whole they seemed to support Einstein and the astronomers on the spot were satisfied in that respect.

The third consequence of relativity was that light, climbing against a gravitational field, would lose some of its energy, that loss being related in a definite way to the intensity of the field. The loss in energy would mean that all the lines in the spectrum of the light would shift slightly toward the red end. This would be an "Einstein red shift," as compared to the better-known "Doppler-Fizeau red shift," which arose when the light source was receding from the observer.

The Einstein red shift was a hard thing to test for. Even the Sun's gravitational field was not intense enough to produce a red shift of this sort large enough to measure.

Then the British astronomer Arthur Stanley Eddington, who was one of the first to accept Einstein's theory wholeheartedly, had an interesting idea. If the Companion of Sirius was of equal mass with the Sun, but had only 1/30 its diameter, then the surface gravity of the Companion ought to be 900 times that of the Sun. The Companion ought, therefore, to subject the light rising from its surface to 900 times the gravitational pull and it might produce an Einstein red shift that could be detected.

Eddington alerted Walter Adams to this fact, since Adams was the world expert on the Companion's spectrum.

Adams got to work. Sirius and its Companion are both receding from us and that produces a red shift, but it does so in both alike, so that was no problem. In addition, Sirius and its Companion circled about each other so that one might be receding relative to the other—but that was a known motion and could be allowed for.

Once all allowances were made, then any residual red shift shown by the Companion's spectrum that was not present in Sirius' had to be an Einstein red shift.

Adams made his measurements carefully and found that there *was* an Einstein red shift and, what's more, that it was exactly as large as Einstein's theory predicted. This was the third and, up to that time, the most unequivocal demonstration of the correctness of the Theory of General Relativity.

It worked the other way around, too. If we assume that

the Theory of General Relativity is correct, then the fact that the Companion shows the Einstein red shift demonstrates conclusively that it must have an extraordinarily high surface gravity and must indeed be far denser than ordinary stars and planets can be.

Since 1922, then, no one has doubted the astonishing characteristics of the Companion and of the other white dwarfs. Yet since then, much more astonishing objects have been found which Adams (who died in Pasadena, California, on May 11, 1956) did not live to see—but that's for Chapter 15.

## Fifteen
# TWINKLE, TWINKLE, MICROWAVES

When I look back over the essays that have appeared in my books and which have been written over the last eighteen and a half years, I'm not too surprised to find an occasional one of them that has become obsolete through the advance of science.

And when that happens, I suppose I am honor-bound, sooner or later, to say so and deal with the matter once again on a newer basis.

Years ago, for instance, I wrote an essay on pygmy stars of various kinds. I entitled it "Squ-u-u-ush," and it appeared in my book *From Earth to Heaven* (Doubleday, 1966).

In it, I discussed, among other things, tiny stars called "neutron stars." I said that there was speculation that one existed in the Crab Nebula, a cloud of very active gas known to be the remnants of a supernova that was seen on Earth just under a thousand years ago. X-rays were given off by the Crab Nebula, and neutron stars might be expected to give off x-rays.

If it were a neutron star, however, the x-rays would be emerging from a point source. The Moon, passing in front of the Crab Nebula, would in that case cut off the x-rays all at once. I went on to say:

"On July 7, 1964, the Moon crossed the Crab Nebula and a rocket was sent up to take measurements . . . Alas, the x-rays cut off gradually. The x-ray source is about a light-year across and is no neutron star.

". . . In early 1965, physicists at C.I.T. recalculated the cooling rate of a neutron star . . . They decided it would . . . radiate x-rays for only a matter of weeks."

The conclusion, apparently, was that it was not very likely that *any* x-ray source could be a neutron star and that these objects, even if they existed, could probably never be detected.

And yet just two years after I wrote the essay (and about eight months after the essay collection was published) neutron stars were discovered after all, and quite a few of them are now known. It's only reasonable that I explain how that came about—by going back a bit.

In the previous chapter I discussed the discovery of white dwarfs.

White dwarfs are stars that have the mass of ordinary stars, but the volume of planets. The first white dwarf to be discovered, Sirius B, has a mass equal to that of our Sun, but a diameter of only 47,000 kilometers (29,900 miles)—about that of Uranus.

How can that be?

A star like the Sun has a sufficiently intense gravitational field to pull its own matter inward with a force that will crush the atoms and reduce them to an electronic fluid within which the much tinier nuclei will move freely. Even if, under those circumstances, the Sun compressed itself to 1/26,000 its present volume and 26,000 times its present density, so that it was a white dwarf the duplicate of Sirius B, it would still be—from the standpoint of the atomic nuclei—mostly empty space.

Yet the Sun does not so compress itself. Why not?

There is nuclear fusion going on at the stellar core which raises the temperature there to about 15,000,000° C. The expansive effect of that temperature balances the inward pull of gravity and keeps the Sun a large ball of incandescent gas with an over-all density of only 1.4 times that of water.

Eventually, however, the nuclear fusion at the center of a star will run out of fuel. This is a complicated process which we don't have to go into here, but in the end there is nothing left to supply the necessary heat at the core—the heat that keeps the star expanded. Gravitation then has its way; there is a stellar collapse, and a white dwarf is formed.

The electronic fluid within which the nuclei of the white dwarf move can be viewed as a kind of spring that resists when it is compressed, and resists more strongly as it is compressed more tightly.

A white dwarf maintains its volume and resists further compression by the gravitational in-pull through this spring action and not by the expansive effect of heat. This means that a white dwarf doesn't have to be hot. It may be hot, to be sure, because of the conversion of gravitational energy into heat in the process of collapse, but this heat can very slowly be radiated away over the eons so that the white dwarf will become, eventually, a "black dwarf." Even so, it will still maintain its volume, the compressed electronic fluid remaining in equilibrium with the gravitational pull forever.

Stars, however, come in different masses. The larger the mass of a star, the more intense its gravitational field. When the nuclear fuel runs out and a star collapses, then the larger its mass and the more intense its gravitational field, the more tightly compressed the white dwarf that results and the smaller.

Eventually, if the star is massive enough, the gravitational pull will be intense enough and the collapse energetic enough to shatter the spring of the electronic fluid, and no white dwarf will then be able to form nor sustain its planetary volume.

An Indian-American astronomer, Subrahmanyan Chandrasekhar, considered the situation, made the necessary calculations, and, in 1931, announced that the shattering would take place if the white dwarf had a mass more than 1.4 times that of the Sun. This mass is called "Chandrasekhar's limit."

Not very many stars have masses beyond that limit— not more than 2 per cent of all the stars in existence do. However, it is precisely the massive stars that run out of nuclear fuel first. The more massive a star, the more quickly it runs out of nuclear fuel and the more drastically it collapses.

Collapse must, in the 15,000,000,000-year life span of the Universe, have taken place to a disproportionate amount among the massive stars. Of all the stars that have consumed their nuclear fuel and collapsed, at least a quarter, possibly more, have had masses greater than Chandrasekhar's limit. What happened to them?

The problem did not bother most asronomers. As a star uses up its nuclear fuel, it expands, and it seems likely that in the ultimate collapse only the inner regions would take part. The outer regions would linger behind to form a "planetary nebula," one in which a bright, collapsed star was surrounded by a vast volume of gas.

To be sure, the mass of the noncollapsed gas of a planetary nebula is not very great, so only stars slightly above the limit, would lose enough mass in this way to be brought safetly below the limit.

On the other hand, there are exploding stars, novas, and supernovas that, in the course of explosion, lose anywhere from 10 to 90 per cent of their total stellar masses. Each explosion spreads dust and gas in all directions, as in the Crab Nebula, leaving only a small inner region, sometimes only a very *small* inner region, to undergo collapse.

One could suppose, then, that whenever the mass of a star was beyond Chandrasekhar's limit, some natural process would remove enough of the mass, to allow whatever portion collapsed to be below Chandrasekhar's limit.

But what if this were not always so? What if we could not trust the benevolence of the Universe that far, and what if sometimes a too-massive conglomeration of matter collapsed?

In 1934 the two American astronomers, Swiss-born Fritz Zwicky and German-born Walter Baade, considered this possibility and decided that the collapsing star would simply crash through the electron-fluid barrier. The electrons, compressed further and further, would be squeezed into the protons of the atomic nuclei moving about in the fluid, and the combination would form neutrons. The main bulk of the star would now consist only of the neutrons present in the nucleus to begin with, plus additional neutrons formed by way of electron-proton combinations.

The collapsing star would thus become virtually nothing but neutrons and it would continue to collapse until the neutrons were essentially in contact. It would then be a "neutron star." If the Sun collapsed into a neutron star, its diameter would be only 1/100,000 what it is now. It would be only 14 kilometers (9 miles) across—but it would retain all its mass.

A couple of years later, the American physicist J. Robert Oppenheimer and a student of his, George M. Volkoff, worked out the theory of neutron stars in detail.

It would appear that white dwarfs were formed when relatively small stars reached their end in a reasonably quiet way. When a massive star exploded in a supernova (as only massive stars do), then the collapse is rapid enough to crash through the electronic-fluid barrier. Even if enough of the star is blown away to leave the collapsing remnant below Chandrasekhar's limit, the speed of collapse may carry it through the barrier. You could therefore end up with a neutron star that was less massive than some white dwarfs.

The question is, though, whether such neutron stars really exist. Theories are all very nice, but unless checked by observation or experiment, they remain only pleasant speculations that amuse scientists and science fiction writers. Well, you can't very well experiment with collapsing stars, and how can you observe an object only a few kilometers across that happens to be at a distance of many light-years?

If you go by light only, it would be difficult indeed, but in forming a neutron star, enough gravitational energy is converted to heat to give the freshly formed object a surface temperature of some 10,000,000° C. That means it would radiate an enormous quantity of very energetic radiation—x-rays, to be exact.

That wouldn't help as far as observers on the Earth's surface were concerned, since x-rays from cosmic sources would not penetrate the atmosphere. Beginning in 1962, however, rockets equipped with instruments designed to detect x-rays were sent beyond the atmosphere. Cosmic x-ray sources were discovered and the question arose as to whether any of them might be neutron stars. By 1965, as I explained in "Squ-u-u-ush," the weight of the evidence seemed to imply they were not.

Meanwhile, however, astronomers were turning more and more to a study of radio-wave sources. In addition to visible light, some of the short-wave radio waves, called "microwaves," could penetrate the atmosphere, and in 1931 an American radio engineer, Karl Jansky, had detected such microwaves coming from the center of the Galaxy.

Very little interest was aroused at the time because astronomers didn't really have appropriate devices for detecting and dealing with such radiation, but during World War II radar was developed. Radar made use of the emission, reflection, and detection of microwaves, and by the end of the war astronomers had a whole spectrum of devices they could now turn to the peaceful use of surveying the heavens.

"Radio astronomy" began and made enormous strides. In fact, astronomers learned how to use complex arrays of microwave-detecting devices ("radio telescopes") that were able to note objects at great distances and with more sharply defined locations than optical telescopes could.

As the technique improved, detection grew finer not only in space but in time. Not only were radio astronomers detecting point sources, but they were also getting indications that the intensity of the waves being emitted could vary with time. In the early 1960s there was even some indication that the variation could be quite rapid, a kind of twinkle.

The radio telescopes weren't designed to handle very rapid fluctuations in intensity because no one had really foreseen the necessity for that. Now special devices were designed that would catch microwave twinkling. In the forefront of this work was the British astronomer Antony Hewish of Cambridge University Observatory. He supervised the construction of 2,048 separate receiving devices spread out in an array that covered an area of 18,000 square meters (or nearly 3 acres).

In July 1967 the new radio telescope was set to scanning the heavens in search of examples of twinkling.

Within a month, a young British graduate student, Jocelyn Bell, who was at the controls of the telescope, was receiving bursts of microwaves from a place midway between the stars Vega and Altair—very rapid bursts, too. In fact, they were so rapid as to be completely unprecedented, and Bell could not believe they came from the sky. She thought she was detecting interference with the radio telescope's workings from electrical devices in the neighborhood. As she went back to the telescope night after night, however, she found the source of the microwaves moving regularly across the sky in time with the stars. Nothing on Earth could be imitating that motion and something in the sky had to be responsible for it. She reported the matter to Hewish.

Both zeroed in on the phenomenon and by the end of November, they were receiving the bursts in such detail that they were able to determine that they were both rapid and regular. Each burst of radio waves lasted only 1/20 of a second and the bursts came at intervals of 1.33 seconds, or about 45 times a minute.

This was not just the detection of a surprising twinkle in a radio source that had already been detected. That particular source had never been reported at all. Earlier radio telescopes were not designed to catch such very brief bursts and would have detected only the average intensity, including the dead period between bursts. The average was only 3 per cent of the maximum burst-intensity, and that went unnoticed.

The regularity of the bursts proved almost unbelievably great. They came so regularly that they could be timed to 1/10,000,000,000 of a second without finding significant variations from pulse to pulse. The period was 1.3370109 seconds.

This was extremely important. If the source were some complex agglomeration of matter—a galaxy, a star cluster, a dust cloud—then parts of it would emit microwaves in a fashion that would differ somewhat from the way other parts did it. Even if each part varied regularly, the meshing together would result in a rather complex resultant. For the microwave bursts detected by Bell and Hewish to be so simple and regular, a very small number of objects, perhaps even a *single* object, had to be involved.

In fact, at first blush, the regularity seemed too much to expect of an inanimate object and there was a slightly scary suspicion that it might represent an artifact after all —but not one in the neighborhood or on Earth. Perhaps these bursts were the extraterrestrial signals some astronomers had been trying to detect. The phenomenon was given the name "LGM" just at first ("little green men").

The LGM notion could not be long maintained, however. The bursts involved total energies perhaps 10,000,-000,000 times that could be produced by all Earth's sources working together, so they represented an enormous investment of energy if they were of intelligent origin. Furthermore, the bursts were so unvaryingly regular that they contained virtually no information. An advanced intelligence would have to be an advanced stupidity to spend so much energy on so little information.

Hewish could only think of the bursts as originating from some cosmic object—a star perhaps—that sent out pulses of microwaves. He therefore called the object a "pulsating star" and that was quickly shortened to "pulsar."

Hewish searched for suspicious signs of twinkles in other places in the records that his instrument had been accumulating, found them, went back to check, and, in due course, was quite sure he had detected three more pulsars. On February 9, 1968, he announced the discovery to the world (and for that discovery eventually received a share of the 1974 Nobel Prize for physics).

Other astronomers around the world began to search the skies avidly and more pulsars were quickly discovered. Over a hundred pulsars are now known, and there may be as many as 100,000 in our Galaxy altogether. The nearest known pulsar may be as close as 300 light-years or so.

All the pulsars are characterized by extreme regularity of pulsation, but the exact period varies from pulsar to pulsar. The one with the longest period has one of 3.75491 seconds (or 16 times a minute).

The pulsar with the shortest period so far known was discovered in October 1968 by astronomers at the National Radio Astronomy Observatory at Green Bank, West Virginia. It happens to be in the Crab Nebula and this was the first clear link between pulsars and supernovas. The Crab Nebula pulsar has a period of only 0.033099 seconds. This is about 1,813 times a minute and is about 113 times as rapid a pulsation as that of the longest-period pulsar known.

But what could produce such rapid and such regular pulsations?

Leaving intelligence out of account, it could only be produced by the very regular movement of one or possibly two objects. These movements could be either (1) the revolution of one object about another with a burst of microwaves at some one point in the revolution; (2) the rotation of a single body about its axis, with a burst at one point in the rotation; or (3) the pulsation, in and out, of a single body, with a burst at one point in the pulsation.

The revolution of one object about another could be that of a planet about its sun. This was the first fugitive thought of the astronomers when the suspicion existed for a while that the bursts were of intelligent origin. However, there is no reasonable way in which a planet could revolve or rotate at a rate that would account for such a rapid regularity in the absence of intelligence.

The fastest revolutions would come when the gravitational fields were most intense and, in 1968, that meant white dwarfs. Suppose you had two white dwarfs, each at the Chandrasekhar limit and revolving about the other in virtual contact. There could be no faster revolution, by 1968 thinking, and that was still not fast enough. The microwave twinkle could not be the result of revolution, therefore.

How about rotation? Suppose a white dwarf were rotating in a period of less than 4 seconds? No go. Even a white dwarf, despite the mighty gravitational field holding it together, would break up and tear apart, if it were rotating that fast—and that went for pulsations as well.

If the microwave twinkle were to be explained at all, what was needed was a gravitational field much more intense than those of white dwarfs—and that left astronomers only one direction in which to go.

The Austrian-born American astronomer Thomas Gold said it first. The pulsars, he suggested, were the neutron stars that Zwicky, Baade, Oppenheimer, and Volkoff had talked about a generation before. Gold pointed out that a neutron star was small enough and had a gravitational field intense enough to be able to rotate about its axis in 4 seconds or less without tearing apart.

What's more, a neutron star should have a magnetic field as any ordinary star might have, but the magnetic field of a neutron star would be as compressed and concentrated as its matter was. For that reason, a neutron star's magnetic field would be enormously more intense than the fields about ordinary stars.

The neutron star, as it whirled on its axis, would give off electrons from its outermost layers (in which protons and electrons would still be existing), thanks to its enormous surface temperature. Those electrons would be trapped by the magnetic field and would be able to escape only at the magnetic poles at opposite sides of the neutron stars.

The magnetic poles would not have to be at the actual rotational poles (they aren't in the case of the Earth, for instance). Each magnetic pole would sweep around the rotational pole in 1 second or in fractions of 1 second and would spray out electrons as it did so (just as a rotating water sprinkler jets out water). As the electrons were thrown off, they would curve in response to the neutron star's magnetic field and lose energy in the process. That energy emerged in the form of microwaves, which were not affected by magnetic fields, and which went streaking off into space.

Every neutron star thus would end by shooting out two jets of radio waves from opposite sides of its tiny globe.

If a neutron star happened to move one of those jets across our line of sight as it rotates, Earth would get a very brief pulse of microwaves at each rotation. Some astronomers estimate that only one neutron star out of a hundred would just happen to send microwaves in our direction, so that of the possibly 100,000 in our Galaxy we might never be able to detect more than 1,000.

Gold went on to point out that, if his theory was correct, the neutron star would be leaking energy at its magnetic poles and its rate of rotation would have to be slowing down. This meant that the faster the period of a pulsar, the younger it was likely to be and the more rapidly it might be losing energy and slowing down.

That fits the fact that the Crab Nebula neutron star is the shortest-period one that is known, since it is not quite a thousand years old and may easily be the youngest we can observe. At the moment of its formation, it might have been rotating 1,000 times a second. The rotation would have slowed rapidly down to a mere 30 times a second now.

The Crab Nebula neutron star was studied carefully and it was indeed found to be lengthening its period. The period is increasing by 36.48 billionths of a second each day and, at that rate, its period of rotation will double in length in 1,200 years. The same phenomenon has been discovered in the other neutron stars whose periods are slower than that of the Crab Nebula and whose rate of rotational slowing is also slower. The first neutron star discovered by Bell, now called CP1919, is slowing its rotation at a rate that will double its period only after 16,000,000 years.

As a pulsar slows its rotation, its bursts of microwaves become less energetic. By the time the period has passed 4 seconds in length, the neutron star would no longer be detectable. Neutron stars probably endure as detectable objects for tens of millions of years, however.

As a result of the studies of the slowing of the microwave bursts, astronomers are now pretty well satisfied that the pulsars are neutron stars, and my old essay "Squ-u-u-ush" stands corrected.

Sometimes, by the way, a neutron star will suddenly speed its period very slightly, then resume the slowing trend. This was first detected in February 1969, when the period of the neutron star Vela X-1 was found to alter suddenly. The sudden shift was called, slangily, a "glitch," from a Yiddish word meaning "to slip," and that word is now part of the scientific vocabulary.

Some astronomers suspect glitches may be the result of a "starquake," a shifting of mass distribution within the neutron star that will result in its shrinking in diameter by 1 centimeter or less. Or perhaps it might be the result of the plunging of a sizable meteor into the neutron star so that the momentum of the meteor is added to that of the star.

There is, of course, no reason why the electrons emerging from a neutron star should lose energy only as microwaves. They should produce waves all along the spectrum. They should, for instance, emit x-rays, too, and the Crab Nebula neutron star does indeed emit them. About 10 to 15 per cent of all the x-rays the Crab Nebula produces is from its neutron star. The other 85 per cent or more, which came from the turbulent gases surrounding the neutron star, obscured this fact and disheartened those astronomers who had hunted for a neutron star there in 1964.

A neutron star should produce flashes of visible light, too.

In January 1969 it was noted that the light of a dim sixteenth-magnitude star within the Crab Nebula *did* flash on and off in precise time with the radio pulses. The flashes were so short and the period between them was so brief that special equipment was required to catch those flashes. Under ordinary observation, the star seemed to have a steady light.

The Crab Nebula neutron star was the first "optical pulsar" discovered, the first *visible* neutron star. (After this essay was first published, a second visible neutron star was detected.)

This doesn't end the story, for my essay "Squ-u-u-ush" was wrong in another and still more spectacular respect than in the case of neutron stars.

Correcting that will enable me to take one more step. In the last chapter we talked about the discovery of that small, dense, stellar monster, the white dwarf.

In this chapter we talked about the discovery of that smaller, denser, stellar supermonster, the neutron star.

Well, in the next chapter it will be time to describe the discovery of that smallest, densest, stellar superest-monster, the black hole.

## Sixteen
# THE FINAL COLLAPSE

When I was young (even younger than I am now, if you can imagine such a thing), I read the books written by my science-writing predecessors.

I was particularly fond of reading about the amazing world of relativity and was much taken with the new geometric view of the Universe—the manner in which space curved in the vicinity of matter, curving more and more sharply as masses were greater and greater and more and more condensed. The gravitational effect, I gathered, was a way of describing the manner in which all objects, even light ones, skidded around the curve.

I was told directly by these books, or I inferred it (I no longer remember which), that if one could get a mass that was large enough and sufficiently condensed, one could imagine space curved so sharply about the body as to leave only a bottleneck connection to the Universe generally. If the mass were even larger and more condensed, the bottleneck would be smaller and smaller until finally, at some critical value of mass and density, it pinched off altogether leaving the supermass effectively isolated in what amounted to a universe of its own and unable in any way to affect the great Universe of which it had once been a part.

Even as late as 1965 I believed this, for in my essay "Squ-u-u-ush," in *From Earth to Heaven* (Doubleday, 1966), after having discussed the neutron star (see also the previous chapter), I went on to discuss an object compressed even more extremely than that. Lacking a name for such an object, I invented the term "superneutron star."

The Sun would become a neutron star if, without losing mass, it were compressed into a tiny sphere 14 kilometers (9 miles) across. If it were compressed still further, into a ball but 6 kilometers (3.7 miles) across, it would become what I called a "superneutron star," with a density and a surface gravity each about 10 times that of a Sun-sized neutron star and an escape velocity equal to the speed of light. Since nothing can go faster than light (excluding the hypothesized, and still problematical, tachyons) nothing—not even light—can leave such a superneutron star.

I visualized the result, in 1965, as being an example of a tiny pinched-off bit of the Universe, unable to affect the rest. I made the following statements about it in "Squ-u-u-ush":

> A superneutron star could not, therefore, affect the rest of the Universe in any way. It could give no sign of its existence; neither radiational nor gravitational. . . .
>
> The superneutron star has been pinched off into a tiny universe all its own, forever closed and self-sufficient. . . .
>
> Naturally, we could never detect a superneutron star even if it existed, no matter how close it was. . . .

Alas, I was wrong. Apparently, while matter and electromagnetic radiation cannot escape from a superneutron star, the gravitational effect will continue to make itself evident. The superneutron star therefore can, and does, affect outside parts of the Universe through its gravitation and does not occupy a Universe of its own. And because it does affect the rest of the Universe, it can, in theory, be detected.

I should also mention that my suggested name "superneutron star" did not catch on with the world in general. Instead, the reasoning went as follows—

A superneutron mass with an escape velocity equal to, or greater than, the speed of light cannot emit particles possessing mass. This means that any piece of ordinary matter can fall in, but can't come out again. The effect is that of falling into an infinitely deep hole in space. What's more, since even light can't emerge for it, we can't see it. It is a completely black hole.

And that's it. The superneutron star is generally known as a "black hole" but I never heard the phrase till after 1965.

Of course, "black hole" doesn't sound very scientific, and the term "collapsed star" has been suggested instead. This would be abbreviated to "collapsar," a form analogous to "quasar" and "pulsar."

I don't think that "collapsar" will ever become the term of choice, however. "Black hole" may sound prosaic, but the picture it gives rise to is so dramatic, and so essentially accurate, that I don't expect it to be abandoned.

So, having talked about white dwarfs in Chapter 14 and neutron stars in Chapter 15, let's go on to black holes.

The mass of a white dwarf, pulled together by an intense gravitational field, is kept from total collapse by the resistance of the electronic fluid, which can be pictured as electrons in contact. Nevertheless, if the mass of a star is too great, it will produce a gravitational field too intense to be held back by the electronic fluid. In that case, the star, when it collapses, will smash right through the white-dwarf stage and collapse to a neutron star, where it is a neutronic fluid, neutrons in contact that withstand further collapse.

Surely, even the mashed-together neutrons must have a limit of resistance. In 1939 J. Robert Oppenheimer reasoned that at some point the neutronic fluid must give way and that when that happened, there existed nothing at all—*nothing*—that could withstand the gravitational collapse. There would be one final collapse to zero volume, and a black hole would form.

It would appear that the crucial mass-level is 3.2 times the Sun's mass, so that there can be no neutron star more massive than that.

About one star in a thousand possesses a mass more than 3.2 times that of our Sun. That doesn't sound like much but it comes to 100,000,000 stars in our Galaxy alone. What's more, these massive stars are short-lived. Whereas our Sun will remain on the main sequence, radiating steadily and quietly, as it does now, for some 12,000,000,000,000 years altogether (5,000,000,000 years of which have elapsed) before it expands and then collapses, these 100,000,000 massive stars will remain on the main sequence for less than 1,000,000,000 years altogether. In the 15,000,000,000-year life span of the Universe, there has been time for many generations of these massive stars to be born and collapse.

The total number of these massive-star collapses may well be in the many billions in the lifetime of our Galaxy. Can all these billions of massive stars have collapsed into black holes?

Not necessarily. Such massive stars will invariably explode as supernovas before collapsing, and the supernova may scatter anywhere up to nine tenths of the mass of a star through space, leaving only a minor remnant to collapse. That minor remnant may be small enough to collapse to a neutron star only.

Can it be, in fact, that a supernova always blows off enough mass to prevent black hole formation? Can it be that every star, no matter how massive, ends up as a neutron star plus a vast cloud of dust and gas?

No, we can't rule out black holes altogether, for it would appear that any star which possesses more than 20 times the mass of the Sun will not be able to get rid of enough mass by supernova explosions to leave less than 3.2 Sun-masses behind. Such a star will collapse to a black hole as a matter of necessity.

There are about 20,000 stars in our Galaxy right now that are of spectral class O and have a mass of anywhere from 20 to 70 times that of the Sun.

Such O-class stars are very short-lived and are not likely to remain on the main sequence for even as long as 1,000,-000 years. During the lifetime of our Universe, we can imagine as many as 15,000 generations of such giant stars coming to birth and eventually collapsing, coming to birth and eventually collapsing—

And, of course, some stars of less than 20 times the Sun's mass might also end up with a collapsing remnant more than 3.2 times the sun's mass.

On the whole, then, we can conclude that there *must* be black holes in the Universe, perhaps even many millions of them in our Galaxy alone.

In that case, if black holes exist, and in reasonable quantities, can they be detected?

You can't detect any particles coming out of them, or any electromagnetic radiation either, but you *can*, in theory, detect gravitational effects.

To be sure, the total gravitational pull exerted by a black hole at a great distance is no greater than the total gravitational pull exerted by its mass in any other form. Thus, if you were 100 light-years away from a giant star with 50 times the mass of the Sun, its gravitational pull would be so diluted by distance that it would be indetectably small. If, somehow, that star were to become a black

—

hole with a mass 50 times the mass of the Sun, its gravitational pull at a distance of 100 light-years would be precisely the same as before and would still be undetectable.

The difference arises at close quarters. The black hole is much smaller in size than a giant star of the same mass. An object near the surface of the black hole is much closer to the center of mass than an object near the surface of the giant star would be. The object near the black hole would experience a gravitational intensity enormously greater than the object near the giant star. (Even if we imagine an object penetrating the surface of the giant star and approaching the center in that fashion, an increasing portion of the mass of the giant star is left behind and the penetrating object is attracted only by the mass closer to the center than it itself is. When the object is within a few kilometers of the star-center, the gravitational pull on it is very small.)

What we can hope to do, then, is to detect not the total gravitational pull of a black hole but the effects of the locally enormous gravitational intensities it produces.

By Einstein's Theory of General Relativity, for instance, gravitational activity releases gravitational waves. These carry so excessively minute an amount of energy that detecting them is virtually beyond hope. If there is any chance of detection at all, this would come about when gravitational waves enormously more energetic than is usual would be produced. To produce *such* gravitational waves, a large black hole in the process of formation or growth might be required.

In the late 1960s the American physicist Joseph Weber used large aluminum cylinders, weighing several tons each and located hundreds of miles apart, as gravitational-wave detectors. Such cylinders would be very slightly compressed and expanded as gravitational waves passed; and since gravitational waves have incredibly long wavelengths, two cylinders, even widely separated, would react simultaneously to the same gravitational wave. In fact, it is this simultaneous reaction that is the surest indication that a gravitational wave is being detected.

Weber reported gravitational waves and produced considerable excitement. Weber's data made it appear that enormously energetic gravitational events were taking place at the center of the Galaxy and that a large black hole might be located there.

Other scientists have, however, tried to repeat Weber's findings and have failed, so that at this time the question of whether gravitational waves have been detected remains moot. There may be a black hole at the center of the Galaxy, but Weber's route to its detection is discounted now and other ways of detecting black holes must be considered.

Another way, still using the black hole's intense gravitational field in its immediate neighborhood, is to study the behavior of light that might be skimming past a black hole. Light will curve slightly in the direction of a gravitational source, and it will do so detectably even when it skims past a large object with an ordinary gravitational field, like our Sun.

Suppose, now, that a black hole is lying precisely between a distant galaxy and Earth. The light of the galaxy will pass the point-like black hole, itself invisible, on all sides. On all sides the light is bent toward the black hole and is made to converge in our direction. The black hole does to light, gravitationally, what a lens does through refraction. The effect, therefore, is spoken of as a "gravitational lens."

If we were to see some galaxy which, despite its distance, looked abnormally large and, possibly, distorted as well, we might suspect it was being magnified by a gravitational lens and that between it and ourselves lies a black hole. No such phenomenon, however, has yet been observed. Some other way of detecting a black hole must be found.

Black holes are not alone in the Universe. There could well be other matter in the vicinity, and such matter, passing near the black hole, might collide with it and be engulfed or might move into orbit about it.

In approaching a black hole, any object larger than a dust particle would be subjected to such enormous tidal forces that it would be reduced to dust. Around the black hole, then, would be an "accretion disc," a kind of asteroidal belt of dust particles about 200 kilometers (125 miles) away from the center.

If the black hole happened to be an isolated one, with no great quantities of matter anywhere within many light-years, the accretion disc would be very thin, perhaps even nonexistent. If, on the other hand, there were a large source of ordinary matter in the immediate neighborhood, a thick and dense accretion disc could result.

We might suppose that an accretion disc would circle the black hole forever, as the Earth circles the Sun. There would be, however, many collisions that pass energy from one particle to another. Some particles would be bound to lose energy and spiral in closer to the black hole. The closer the spiral, the harder it would be for a particle to back out again, and once it got past a certain critical difference, it could nevermore emerge.

There would thus be a continual drizzle of matter entering the black hole. Nor would the accretion disc necessarily die out, since further supplies of matter would continue to arrive from whatever matter-source existed in the neighborhood.

Matter entering the black hole would lose gravitational energy, which would be converted into heat. The matter would be further heated by the stretching and compression of tidal forces. The result would be that matter entering the black hole would be heated to enormous temperatures and give off a whole range of electromagnetic radiation, right up to energetic x-rays.

Thus, while we cannot detect a bare black hole surrounded by utter vacuum, we might conceivably detect one that is swallowing matter, since that matter would, as its death cry, emit x-rays.

The x-rays would have to be intense enough to detect across many light-years of space, so they would have to represent more than a thin drizzle of occasional dust. There would have to be torrents of matter swirling inward and this would mean that the black hole, to be detected, would have to be in pretty specialized surroundings. It would have to be within easy reach of large supplies of matter.

Those regions where stars are thickly strewn are therefore much more likely to have a detectable black hole than are those regions where stars are sparse. Nowhere are stars distributed more thickly than at the cores of galaxies and it is there, perhaps, that we ought to look.

In recent years there has been increasing evidence that some spectacularly energetic explosions have taken place in galactic cores in the past and, in a few cases, even as we watch. Could black holes in some form or other be responsible?

Indeed, a very compact and energetic microwave source has been detected at the center of our own Galaxy. Could that represent a black hole there? Some astronomers speculate that this is so and that our galactic black hole has the mass of 100,000,000 stars, or 1/1,000 the entire Galaxy. It would have a diameter of 700,000,000 kilometers (43,-000,000 miles) and be large enough to disrupt whole stars through tidal effects or gulp them down whole before they could break up, if the approach was fast enough.

Perhaps every galaxy has a black hole at the core, and if that is so, then the one of that type that is nearest to ourselves is the one in our own Galaxy, of course, and it is 30,000 light-years away. A large black hole would be an uncomfortable near neighbor, but 30,000 light-years is satisfactory insulation.

It may be that galactic cores are not the only places where detectable black holes occur. Outside the core there are globular clusters made up of tens of thousands or even hundreds of thousands of stars packed together in a spherical conglomerate.

Such globular clusters (of which there are a couple of hundred in our Galaxy) are not as densely packed as a galactic core is, but at their centers, star concentrations are very much higher than they are in our Sun's neighborhood, for instance.

And, as a matter of fact, a number of globular clusters have been tabbed as x-ray sources. The possibility therefore exists that there are indeed black holes at the center of some clusters, and perhaps of every cluster. Some astronomers speculate that such globular-cluster black holes may have masses 10 to 100 times that of the Sun.

In that case, there are some detectable black holes closer than the core of our Galaxy. The closest globular-cluster black hole would be the one in Omega Centauri, which is 22,000 light-years away.

The trouble with black holes at the center of galaxies or of globular clusters is that you can't get a real look at them. You can detect unusual radiation and infer that there *may* be a black hole there. However, since many thousands or even millions of ordinary stars lie between the possible black hole and ourselves, forming an impenetrable barrier to any closer examination, we can only guess as to the black hole's existence, and we may never be able to do more than that.

What we need, then, is a black hole that has plenty of matter in its vicinity, enough to form a large accretion disc, and yet one that is sufficiently alone in space so that we can study the spot where it is located without anything of importance in between.

This might seem like a mutually exclusive pair of requirements, but it isn't. What we want is a binary—a pair of stars revolving about a mutual center of gravity, with one of them a black hole and one a normal star. In that case, if the two objects are close enough, there may be a sufficient leak of matter from the normal star to the black hole to form an accretion disc, which would, in turn, serve as an x-ray source.

So we should look for some object in the heavens that consists of a normal star and an x-ray source circling about a mutual center of gravity, with no star visible at the site of the x-ray source.

In the early 1960s, x-ray sources were first discovered in the sky through the use of detectors carried beyond our atmosphere by rockets (X-rays will not penetrate our atmosphere.) In 1965 a particularly intense x-ray source was detected in the constellation Cygnus and was named Cygnus X-1. It is thought to be about 10,000 light-years from us.

The mere fact that Cygnus X-1 was so intense an x-ray source roused interest. In those years, neutron stars were still being searched for and it was thought that Cygnus X-1 might well be a neutron star.

In 1970 an x-ray-detecting satellite was launched from the coast of Kenya on the seventh anniversary of Kenyan independence. It was named "Uhuru" from the Swahili word for "freedom." It multiplied knowledge of x-ray sources to unlooked-for heights, since it detected 161 of them. Half the x-ray sources it detected were in our own Galaxy and three of them were in globular clusters.

In 1971 Uhuru detected a marked change in x-ray intensity in Cygnus X-1 and this was particularly interesting. By that time neutron stars had been discovered and it was known that x-rays emitted by them would come in regular pulses. An irregular change was much more likely to arise from a black hole, where much would depend on events in the accretion disc and where matter might push in in greater quantities at some times than others. The fact that Uhuru detected that sudden change in Cygnus X-1, combined with its x-ray brilliance, made the chance of a black hole look suddenly good.

It was necessary to locate Cygnus X-1 with great exactness and that could best be done by microwaves, which should also be arriving from the source if it were a black hole. Microwaves were indeed detected and the use of sophisticated radio telescopes made it possible to pinpoint the source very accurately and place it just next to a visible star.

The star was HD-226868, a large, hot, blue star of spectral class B, which was some 30 times as massive as our Sun. An astronomer at the University of Toronto, C. T. Bolt, showed HD-226868 to be a binary. It is clearly circling in an orbit with a period of 5.6 days—an orbit, the nature of which makes it appear that the other object in the binary is perhaps 5 to 8 times as massive as the Sun.

The companion star cannot be seen, however, so it cannot be a normal star. A normal star with a mass 5 to 8 times that of the Sun would be dimmer than its companion, but it would be bright enough to see. The only reason it would not be seen would be its small size. It might be a white dwarf, a neutron star, or a black hole. A white dwarf can't be more massive than 1.4 times the Sun, and a neutron star can't be more massive than 3.2 times the Sun. That leaves us with only a black hole as a possibility, and one that is closer than any galactic-core black hole or globular-cluster black hole could possibly be.

Another point in favor of the black hole hypothesis is that HD-226868 seems to be expanding as though it were entering its red-giant stage. Its matter would therefore very likely be spilling over into the grip of the black-hole companion. This would form a large accretion disc and would account for the fact that Cygnus X-1 is such an intense x-ray source.

The one catch is the matter of distance.

Suppose you have a particular binary under study and can establish a carefully observed angular separation and period. That angular separation of so many hundredths of a second of arc can be converted into a spatial separation of so many millions of kilometers (or miles), if you know the distance. The greater the distance, the greater the real separation must be to produce the observed angular separation.

But the greater the real separation, the greater the gravitational interaction between the stars must be to produce the observed period. The greater the gravitational interaction between the stars, the greater the total mass of the two stars.

If Cygnus X-1 is indeed 10,000 light-years away, then the mass of the two stars is as I've given them and the X-ray source is too massive to be anything but a black hole.

If Cygnus X-1 is, however, for some reason, much closer than that (and stellar distances, except for the nearest stars, can be quite uncertain), then the mass of the stars would be much less than we think they are. In that case, the invisible object serving as an x-ray source might be a neutron star or a white dwarf, rather than a black hole. Some have even suggested that it could, conceivably, be that utterly unremarkable object, a red dwarf. (Three quarters of all the stars in existence are red dwarfs.)

However, the astronomical majority seems to hold for the 10,000 light-year distance and the black-hole—and since black holes are dramatic and imagination-stretching, that is a pleasant thought to science fiction individuals like myself.

# VII

---

# OURSELVES

# Seventeen
# ASIMOV'S COROLLARY

I have just come back from Rensselaerville, New York, where, for the fifth year, I have led a four-day seminar on some futuristic topics. (This time it was on the colonization of space.) Some seventy to eighty people attended, almost all of them interested in science fiction and all of them eager to apply their imaginations to the posing of problems and the suggesting of solutions.

The seminar only runs from a Sunday to a Thursday, but by Thursday there is mass heartbreak at the thought of leaving and vast promises (usually kept) to return the next year.

This year we managed to persuade Ben Bova (editor of *Analog*) and his charming wife, Barbara, to attend. They threw themselves into the sessions with a will and were beloved by all.

Finally came the end, at Thursday noon, and, as is customary on these occasions, I was given a fancy pseudo-plaque testifying to my good nature and to my suave approach toward members of the opposite sex.*

A charming young woman, not quite five feet tall, made the presentation and in simple gratitude, I placed my arm about her waist. Owing to her unusually short height, however, I didn't manage to get low enough and the result brought laughter from the audience.

Trying to dismiss this embarrassing *faux pas* (though I must admit that neither of us budged), I said, "I'm sorry, folks. That's just the Asimov grip."

And from the audience Ben Bova (who, it seems appropriate to say in this particular connection, is my bosom buddy) called out, "Is that anything like the swine flu?"

I was wiped out, and what does one do when one has

---

* See my book *The Sensuous Dirty Old Man* (Walker, 1971).

been wiped out by a beloved pal? Why, one turns about and proceeds to try to wipe out some other beloved pal. — In this case, my English colleague Arthur C. Clarke.

In Arthur's book *Profiles of the Future* (Harper & Row, 1962) he advances what he himself calls "Clarke's Law." It goes as follows:

> *When a distinguished but elderly scientist states that something is possible, he is almost certainly right. When he states that something is impossible, he is very probably wrong.*

Arthur goes on to explain what he means by "elderly." He says: "In physics, mathematics, and astronautics it means over thirty; in the other disciplines, senile decay is sometimes postponed to the forties."

Arthur goes on to give examples of "distinguished but elderly scientists" who have pished and tut-tutted all sorts of things that have come to pass almost immediately. The distinguished Briton Ernest Rutherford pooh-poohed the possibility of nuclear power, the distinguished American Vannevar Bush bah-humbugged intercontinental ballistic missiles, and so on.

But naturally when *I* read a paragraph like that, knowing Arthur as I do, I begin to wonder if, among all the others, he is thinking of me.

After all, I'm a scientist. I am not exactly a "distinguished" one but nonscientists have gotten the notion somewhere that I am, and I am far too polite a person to subject them to the pain of disillusionment, so I don't deny it. And then, finally, I am a little over thirty and have been a little over thirty for a long time, so I qualify as "elderly" by Arthur's definition. (So does he, by the way, for he is—ha, ha—three years older than I am.)

Well, then, as a distinguished but elderly scientist, have I been going around stating that something is impossible or, in any case, that that something bears no relationship to reality? Heavens, yes! In fact, I am rarely content to say something is "wrong" and let it go at that. I make free use of terms and phrases like "nonsense," "claptrap," "stupid folly," "sheer idiocy," and many other bits of gentle and loving language.

Among currently popular aberrations, I have belabored without stint Velikovskianism, astrology, flying saucers, and so on.

While I haven't yet had occasion to treat these matters in detail, I also consider the views of the Swiss Erich von Däniken on "ancient astronauts" to be utter hogwash; I take a similar attitude to the widely held conviction (reported, but not to my knowledge subscribed to, by Charles Berlitz in *The Bermuda Triangle*) that the "Bermula triangle" is the hunting ground of some alien intelligence.

Doesn't Clarke's Law make me uneasy, then? Don't I feel as though I am sure to be quoted extensively, and with derision, in some book written a century hence by some successor to Arthur?

No, I don't. Although I accept Clarke's Law and think Arthur is right in his suspicion that the forward-looking pioneers of today are the backward-yearning conservatives of tomorrow,† I have no worries about myself. I am very selective about the scientific heresies I denounce, for I am guided by what I call Asimov's Corollary to Clarke's Law. Here is Asimov's Corollary:

> *When, however, the lay public rallies round an idea that is denounced by distinguished but elderly scientists and supports that idea with great fervor and emotion —the distinguished but elderly scientists are then, after all, probably right.*

But why should this be? Why should I, who am not an elitist, but an old-fashioned liberal and an egalitarian (see "Thinking About Thinking," in *The Planet That Wasn't*, Doubleday, 1976), thus proclaim the infallibility of the majority, holding it to be infallibly wrong?

The answer is that human beings have the habit (a bad one, perhaps, but an unavoidable one) of being human; which is to say that they believe in that which comforts them.

For instance, there are a great many inconveniences and disadvantages to the Universe as it exists. As examples:

† Heck, Einstein himself found he could not accept the uncertainty principle and, in consequence, spent the last thirty years of his life as a living monument and nothing more. Physics went on without him.

you cannot live forever, you can't get something for nothing, you can't play with knives without cutting yourself, you can't win every time, and so on and so on (see "Knock Plastic," in *Science, Numbers, and I*, Doubleday, 1968).

Naturally, then, anything which promises to remove these inconveniences and disadvantages will be eagerly believed. The inconveniences and disadvantages remain, of course, but what of that?

To take the greatest, most universal, and most unavoidable inconvenience, consider death. Tell people that death does not exist and they will believe you and sob with gratitude at the good news. Take a census and find out how many human beings believe in life after death, in heaven, in the doctrines of spiritualism, in the transmigration of souls. I am quite confident you will find a healthy majority, even an overwhelming one, in favor of side-stepping death by believing in its nonexistence through one strategy or another.

Yet as far as I know, there is not one piece of evidence ever advanced that would offer any hope that death is anything other than the permanent dissolution of the personality and that beyond it, as far as individual consciousness is concerned, there is nothing.

If you want to argue the point, present the evidence. I must warn you, though, that there are some arguments I won't accept.

I won't accept any argument from authority. ("The Bible says so.")

I won't accept any argument from internal conviction. ("I have faith it's so.")

I won't accept any argument from personal abuse. ("What are you, an atheist?")

I won't accept any argument from irrelevance. ("Do you think you have been put on this Earth just to exist for a moment of time?")

I won't accept any argument from anecdote. ("My cousin has a friend who went to a medium and talked to her dead husband.")

And when all that (and other varieties of nonevidence) are eliminated, there turns out to be nothing.‡

‡ Lately, there have been detailed reports about what people are supposed to have seen during "clinical death."—I don't believe a word of it.

Then why do people believe? Because they want to. Because the mass desire to believe creates a social pressure that is difficult (and, in most times and places, dangerous) to face down. Because few people have had the chance of being educated into the understanding of what is meant by evidence or into the techniques of arguing rationally.

But mostly because they want to. And that is why a manufacturer of toothpaste finds it insufficient to tell you that it will clean your teeth almost as well as the bare brush will. Instead he makes it clear to you, more or less by indirection, that his particular brand will get you a very desirable sex partner. People, wanting sex somewhat more intensely than they want clean teeth, will be the readier to believe.

Then, too, people generally love to believe the dramatic, and incredibility is no bar to the belief but is, rather, a positive help.

Surely we all know this in an age when whole nations can be made to believe in any particular bit of foolishness that suits their rulers and can be made willing to die for it, too. (This age differs from previous ages in this, however, only in that the improvement of communications makes it possible to spread folly with much greater speed and efficiency.)

Considering their love of the dramatic, is it any surprise that millions are willing to believe, on mere say-so and nothing more, than alien spaceships are buzzing around the Earth and that there is a vast conspiracy of silence on the part of government and scientists to hide that fact? No one has ever explained what government and scientists hope to gain by such a conspiracy or how it can be maintained, when every other secret is exposed at once in all its details—but what of that? People are always willing to believe in any conspiracy on any subject.

People are also willing and eager to believe in such dramatic matters as the supposed ability to carry on intelligent conversations with plants, the supposed mysterious force that is gobbling up ships and planes in a particular part of the ocean, the supposed penchant of Earth and Mars to play Ping-Pong with Venus and the supposed accurate description of the result in the Book of Exodus, the supposed excitement of visits from extraterrestrial astronauts in prehistoric times and their donation to us of our arts, techniques, and even some of our genes.

To make matters still more exciting, people like to *feel* themselves to be rebels against some powerful repressive force—as long as they are sure it is quite safe. To rebel against a powerful political, economic, religious, or social establishment is very dangerous and very few people dare do it, except, sometimes, as an anonymous part of a mob. To rebel against the "scientific establishment," however, is the easiest thing in the world, and anyone can do it and feel enormously brave, without risking as much as a hangnail.*

Thus, the vast majority, who believe in astrology and think that the planets have nothing better to do than form a code that will tell them whether tomorrow is a good day to close a business deal or not, become all the more excited and enthusiastic about the bilge when a group of astronomers denounce it.

Again, when a few astronomers denounced the Russian-born American Immanuel Velikovsky, they lent the man (and, by reflection, his followers) an aura of the martyr, which he (and they) assiduously cultivate, though no martyr in the world has ever been harmed so little or helped so much by the denunciations.

I used to think, indeed, that it was entirely the scientific denunciations that had put Velikovsky over the top and that had the American astronomer Harlow Shapley only had the *sang froid* to ignore the Velikovskian folly, it would quickly have died a natural death.

I no longer think so. I now have greater faith in the bottomless bag of credulity that human beings carry on their back. After all, consider Von Däniken and his ancient astronauts. Von Däniken's books are even less sensible than Velikovsky's and are written far more poorly,† and yet he does well. What's more, no scientist, as far as I know, has deigned to take notice of Von Däniken. Per-

---

* A reader once wrote me to say that the scientific establishment could keep you from getting grants, promotions, and prestige, could destroy your career, and so on. That's true enough. Of course, that's not as bad as burning you at the stake or throwing you in a concentration camp, which is what a *real* establishment could and would do, but even depriving you of an appointment is rotten. However, that works only if you are a scientist. If you are a nonscientist, the scientific establishment can do nothing more than make faces at you.

† Velikovsky, to do him justice, is a fascinating writer and has an aura of scholarliness than Von Däniken utterly lacks.

haps they felt such notice would do him too much honor and would but do for him what it had done for Velikovsky.

So Van Däniken has been ignored—and, despite that, is even *more* successful than Velikovsky is, attracts more interest, and makes more money.

You see, then, how I choose my "impossibles." I decide that certain heresies are ridiculous and unworthy of any credit not so much because the world of science says, "It is not so!" but because the world of nonscience says, "It is," so enthusiastically. It is not so much that I have confidence in scientists being right, but that I have so much in nonscientists being wrong.

I admit, by the way, that my confidence in scientists being right is somewhat weak. Scientists have been wrong, even egregiously wrong, many times. There have been heretics who have flouted the scientific establishment and have been persecuted therefor (as far as the scientific establishment is able to persecute), and, in the end, it has been the heretic who has proved right. This has happened not only once, I repeat, but many times.

Yet that doesn't shake the confidence with which I denounce those heresies I do denounce, for in the cases in which heretics have won out, the public has, almost always, not been involved.

When something new in science is introduced, when it shakes the structure, when it must in the end be accepted, it is usually something that excites scientists, sure enough, but does not excite the general public—except perhaps to get them to yell for the blood of the heretic.

Consider Galileo, to begin with, since he is the patron saint (poor man!) of all self-pitying crackpots. To be sure, he was not persecuted primarily by scientists for his scientific "errors," but by theologians for his very real heresies (and they were real enough by seventeenth-century standards).

Well, do you suppose the general public supported Galileo? Of course not. There was no outcry in his favor. There was no great movement in favor of the Earth going round the Sun. There were no "sun-is-center" movements denouncing the authorities and accusing them of a conspiracy to hide the truth. If Galileo had been burned at the stake, as Giordano Bruno had been a generation earlier, the action would probably have proved popular with those parts of the public that took the pains to notice it in the first place.

Or consider the most astonishing case of scientific heresy since Galileo—the matter of the English naturalist Charles Robert Darwin. Darwin collected the evidence in favor of the evolution of species by natural selection and did it carefully and painstakingly over the decades, then published a meticulously reasoned book that established the fact of evolution to the point where no rational biologist can deny it‡ even though there are arguments over the details of the mechanism.

Well, then, do you suppose the general public came to the support of Darwin and his dramatic theory? They certainly knew about it. His theory made as much of a splash in his day as Velikovsky did a century later. It was certainly dramatic—imagine species developing by sheer random mutation and selection, and human beings developing from apelike creatures! Nothing any science fiction writer ever dreamed up was as shatteringly astonishing as that to people who from earliest childhood had taken it for established and absolute truth that God had created all the species ready-made in the space of a few days and that man in particular was created in the divine image.

Do you suppose the general public supported Darwin and waxed enthusiastic about him and made him rich and renowned and denounced the scientific establishment for persecuting him? You know they didn't. What support Darwin did get was from scientists. (The support any rational scientific heretic gets is from scientists, though usually from only a minority of them at first.)

In fact, not only was the general public against Darwin then, they are against Darwin now. It is my suspicion that if a vote were taken in the United States right now on the question of whether man was created all at once out of the dirt or through the subtle mechanisms of mutation and natural selection over millions of years, there would be a large majority who would vote for the dirt.

There are other cases, less famous, where the general public didn't join the persecutors only because they never heard there was an argument.

In the 1830s the greatest chemist alive was the Swede

‡ Please don't write to tell me that there are creationists who call themselves biologists. Anyone can call himself a biologist.

Jöns Jakob Berzelius. Berzelius had a theory of the structure of organic compounds which was based on the evidence available at that time. The French chemist August Laurent collected additional evidence that showed that Berzelius' theory was inadequate. He himself suggested an alternate theory of his own which was more nearly correct and which, in its essentials, is still in force now.

Berzelius, who was in his old age and very conservative, was unable to accept the new theory. He retaliated furiously and none of the established chemists of the day had the nerve to stand up against the great Swede.

Laurent stuck to his guns and continued to accumulate evidence. For this he was rewarded by being barred from the more famous laboratories and being forced to remain in the provinces. He is supposed to have contracted tuberculosis as a result of working in poorly heated laboratories and he died in 1853 at the age of forty-six.

With both Laurent and Berzelius dead, Laurent's new theory began to gain ground. In fact, one French chemist who had originally supported Laurent but had backed away in the face of Berzelius' displeasure now accepted it again and actually tried to make it appear that it was *his* theory. (Scientists are human, too.)

That's not even a record for sadness. The German physicist Julius Robert Mayer, for his championship of the law of conservation of energy in the 1840s, was driven to madness. Ludwig Boltzmann, the Austrian physicist, for his work on the kinetic theory of gases in the late nineteenth century, was driven to suicide. The work of both is now accepted and praised beyond measure.

But what did the public have to do with all these cases? Why, nothing. They never heard of them. They never cared. It didn't touch any of their great concerns. In fact, if I wanted to be completely cynical, I would say that the heretics were in this case right and that the public, somehow sensing this, yawned.

This sort of thing goes on in the twentieth century, too. In 1912 a German geologist, Alfred Lothar Wegener, presented to the world his views on continental drift. He thought the continents all formed a single lump of land to begin with and that this lump, which he called "Pangaea," had split up and that the various portions had drifted apart. He suggested that the land floated on the soft, semisolid underlying rock and that the continental pieces drifted apart as they floated.

Unfortunately, the evidence seemed to suggest that the underlying rock was far too stiff for continents to drift through and Wegener's notions were dismissed and even hooted at. For half a century the few people who supported Wegener's notions had difficulty in getting academic appointments.

Then, after World War II, new techniques of exploration of the sea bottom uncovered the global rift, the phenomenon of sea-floor spreading, the existence of crustal plates, and it became obvious that the Earth's crust *was* a group of large pieces that were constantly on the move and that the continents were carried with the pieces. Continental drift, or "plate tectonics," as it is more properly called, became the cornerstone of geology.

I personally witnessed this turnabout. In the first two editions of my *Guide to Science* (Basic Books, 1960, 1965), I mentioned continental drift but dismissed it haughtily in a paragraph. In the third edition (1972) I devoted several pages to it and admitted having been wrong to dismiss it so readily. (This is no disgrace, actually. If you follow the evidence you *must* change as additional evidence arrives and invalidates earlier conclusions. It is those who support ideas for emotional reasons only who can't change. Additional evidence has no effect on emotion.)

If Wegener had not been a true scientist, he could have made himself famous and wealthy. All he had to do was to take the concept of continental drift and bring it down to Earth by having it explain the miracles of the Bible. The splitting of Pangaea might have been the cause, or the result, of Noah's Flood. The formation of the Great African Rift might have drowned Sodom. The Israelites crossed the Red Sea because it was only a half mile wide in those days. If he had said all that, the book would have been eaten up and he could have retired on his royalties.

In fact, if any reader wants to do this *now*, he can still get rich. Anyone pointing out this article as the inspirer of the book will be disregarded by the mass of "true believers," I assure you.

So here's a new version of Asimov's Corollary, which you can use as your guide in deciding what to believe and what to dismiss:

*If a scientific heresy is ignored or denounced by the general public, there is a chance it may be right. If a scientific heresy is emotionally supported by the general public, it is almost certainly wrong.*

You'll notice that in my two versions of Asimov's Corollary I was careful to hedge a bit. In the first I say that scientists are "probably right." In the second I say that the public is "almost certainly wrong." I am not absolute. I hint at exceptions.

Alas, not only are people human; not only are scientists human; but I'm human, too. I want the Universe to be as *I* want it to be and that means completely logical. I want silly, emotional judgments to be *always* wrong.

Unfortunately, I can't have the Universe the way I want it, and one of the things that makes me a rational being is that I know this.

Somewhere in history there are bound to be cases in which science said "No" and the general public, for utterly emotional reasons, said "Yes" and in which it was the general public that was right. I thought about it and came up with an example in half a minute.

In 1798 the English physician Edward Jenner, guided by old wives' tales based on the kind of anecdotal evidence I despise, tested to see whether the mild disease of cowpox did indeed confer immunity upon humans from the deadly and dreaded disease of smallpox. (He wasn't content with the anecdotal evidence, you understand; he *experimented.*) Jenner found the old wives were correct and he established the technique of vaccination.

The medical establishment of the day reacted to the new technique with the greatest suspicion. Had it been left to them, the technique might well have been buried.

However, popular acceptance of vaccination was immediate and overwhelming. The technique spread to all parts of Europe. The British royal family was vaccinated; the British Parliament voted Jenner ten thousand pounds. In fact, Jenner was given semidivine status.

There's no problem in seeing why. Smallpox was an unbelievably frightening disease, for when it did not kill, it permanently disfigured. The general public therefore was almost hysterical with desire for the truth of the suggestion that the disease could be avoided by the mere prick of a needle.

And in this case, the public was right! The Universe *was* as they wanted it to be. In eighteen months after the introduction of vaccination, for instance, the number of deaths from smallpox in England was reduced to one third of what it had been.

So there are indeed exceptions. The popular fancy is sometimes right.

But not often, and I must warn you that I lose no sleep over the possibility that any of the popular enthusiasms of today are liable to turn out to be scientifically correct. Not an hour of sleep do I lose; not a minute.

# DISCUS BOOKS
### DISTINGUISHED NON-FICTION

## A SELECTION OF RECENT TITLES

DRT 2-78